Beyond the Nest: How Biotic Factors Influence Breeding Success

Parkar

Table of Contents

Introduction

Habitat quality and its effect on habitat selection and use

In heterogeneous environments, animals face decisions about how to use habitats of varying quality (Dall et al. 2005). An important aspect of habitat selection is that individuals assess habitat quality correctly in order to select a habitat that maximizes their fitness (Morris 2003). During the breeding or nesting season, these decisions are especially important because they may affect offspring's fitness (Schmidt et al. 2010). However, breeding site selection does not always maximize offspring fitness (Mayhew 1997). When multiple habitats are present to choose from, costs and constraints must be accounted for by an animal when determining if it should use a certain habitat over another (Acker et al. 2017), and this can be especially true for selection among nesting and breeding habitats (Lima 2009, Buxton and Sperry 2017).

Biotic conditions can affect aspects of habitat selection and overall habitat quality, which in turn can affect individuals, populations, and communities. Biotic habitat quality cues can include the presence of conspecifics, heterospecific competitors, and predators, which each can affect the perceived quality of a given habitat for organisms ranging from invertebrates to anurans and birds (Resetarits and Wilbur 1991, Rieger et al. 2004, Fontaine and Martin 2006, Silberbush and Resetarits 2017). These biotic conditions can be especially important at breeding sites and can thus influence how organisms select and use habitats during the breeding season. Males and females, however, may use similar or different cues to assess habitat quality when selecting a breeding location (Resetarits and Wilbur 1991, Eterovick and Ferreira 2008). Whether or not male and female responses to habitat quality are similar and coordinated with one another or different and in conflict with one another is important for understanding why individuals use the habitats they do during the breeding season. Linking male and female breeding habitat preference can then have important implications for both population dynamics and broader community structure on the landscape.

Abiotic conditions are also an important component of overall habitat quality. In both the aquatic and terrestrial environments, these conditions can be altered by anthropogenic activities. An important abiotic characteristic of forest habitats in Europe and North America that has been altered by anthropogenic activities is soil pH. Acid deposition resulting from anthropogenic activities, predominantly the combustion of fossil fuels since the Industrial Revolution, has led to acidification of forest soils (Burns et al. 2011). Acidification can have broad effects on forest floral composition and health, negatively affecting a diverse array or organisms including soil biota (Kuperman et al. 2002), mammals (Pabian et al. 2012), and birds (Hames et al. 2002). Although acidified habitats may directly affect organisms, effects may also occur indirectly due to changes in interspecific interactions and trophic dynamics (Graveland et al. 1994).

Amphibians as a model system in habitat selection and use studies

Amphibians are an excellent system for studying the effects biotic and abiotic conditions on habitat selection and use. Amphibians have complex life histories, often involving an aquatic larval stage and a terrestrial adult stage, so they utilize both aquatic and terrestrial environments. Additionally, because most larvae develop in aquatic habitats with no parental care, the conditions of the environment where the parents breed and deposit their offspring is the environment the offspring will have to develop in.

Therefore, selective pressures should exist among reproducing individuals to discriminate between high- and low-quality breeding sites.

The eastern gray treefrog (*Hyla versicolor*) is a common North American treefrog with a breeding season that occurs from May through July at my study site in Ohio, USA. Males typically arrive at breeding sites before females and establish calling locations to attract females. Females then visit the breeding sites and select a mate among the calling males. Treefrogs are a well-studied group, especially as it relates to habitat selection (Resetarits 1996). This makes them a great system to study both how habitat quality may affect their use of breeding habitats and also how individuals may use these breeding habitats differently based on both the characteristics of the habitat and the characteristics of the individual. Breeding treefrogs use a variety of aquatic habitats (ephemeral to permanent) over an extended breeding season that can be characterized by varying biotic and abiotic conditions. A variety of biotic factors can negatively influence breeding habitat selection, including the presence of certain predators (Rieger et al. 2004) and older conspecific competitors (Resetarits and Wilbur 1989). However, relatively few studies have tested if treefrogs prefer to breed in habitats based on the presence of heterospecific anurans, and these studies have produced mixed results including attraction to, avoidance of, and no preference (Resetarits and Wilbur 1989, Indermaur et al. 2010, Dananay and Benard 2018). Furthermore, few studies have linked breeding habitat preference to offspring performance among habitats containing heterospecific anurans. The conspicuous advertisement call of male treefrogs also made them an excellent study organism for my research. Females prefer longer calls to shorter calls as well as males spending more time actively calling (Schwartz et al. 2001) because these males produce offspring that perform better as tadpoles (Welch et al. 1998). This allowed me to relate the quality of a male (based on these call characters) to the quality of the habitats he chose during the breeding season (based on the characteristics of the habitat).

The American toad (*Anaxyrus americanus*) is another common and well-studied anuran species. As a terrestrial juvenile and adult, American toads are habitat generalists that use a variety of habitat types and thus habitats of different quality. Forests are an important habitat for American toads, and these habitats can be characterized by a variety of biotic and abiotic conditions, including variation in soil properties such as pH. Soil pH can negatively affect amphibians (Wyman and Hawksley-Lescault 1987), but some populations are locally adapted to low pH (Bondi et al. 2016). Therefore, American toads may experience different quality habitats depending on the habitats they use in the terrestrial environments, and these habitats can then affect the American toad at the individual and population level, as well as the broader forest community it is a member of.

Research program and goals

Using a combination of outdoor experimental studies in both aquatic and terrestrial mesocosms, I investigated how biotic and abiotic characteristics affected habitat selection and the effects of using those habitats on two common North American amphibians. My goal was to better understand how and why amphibians use the habitats they do and what the consequences are of using those habitats. I was especially interested in examining these consequences among different life stages of these amphibians.

I first investigated how female breeding habitat selection is influenced by competitors and if this habitat selection behavior is adaptive for offspring performance.

Using an outdoor choice experiment coupled with a mesocosm study, I found that female gray treefrogs used pools with no heterospecific competitors more frequently and earlier in the breeding season than pools with heterospecifics. Tadpoles raised in the presence and absence of these heterospecifics demonstrated that female preference is adaptive, as tadpoles grew larger and developed faster in the absence of heterospecifics. This study fills an important gap in the current literature examining how nonpredatory heterospecific competitors affect the female preference-offspring performance theory in habitat selection.

Because males may use habitat quality cues differently than females, I also evaluated male eastern gray treefrog use of these same habitats during the same outdoor choice experiment. I also evaluated the quality of these males as potential offspring sires for females using recordings of the advertisement calls made by these males. Males used pools with heterospecifics the same as pools without, and the advertisement calls of males at each of the three habitat types were similar. Thus, male and female use of cues from green frog and bullfrog tadpoles at a breeding habitat are not completely coupled. Instead, males may space themselves out amongst multiple available habitats regardless of the larval competitor community present in order to minimize the competition they face from other neighboring advertising males.

The presence of fringe habitats (i.e., those on the periphery of or adjacent to core habitats) on a landscape provide additional habitat choices for animals actively seeking out a breeding location. In Chapters 1 and 2, the pools I established that were used by males and females would be described as fringe habitats, so I wanted to test how and why individuals using these fringe pools may differ from males at an adjacent core breeding pond. Specifically, I expanded my research on the use of these habitats by male treefrogs to test if male treefrogs using fringe habitats differ from residents at an adjacent core pond. I used a combination of observational approaches and an outdoor choice experiment and found that fringe males produced more calls yet shorter calls than males in the core habitat but called at a similar effort. Using a combination of observational approaches and experimental manipulations, I also evaluated how fringe males changed their calls at the core pond. When fringe males were moved to the core habitat, they adjusted their calls to match the core males by lengthening their calls and reducing the overall number of calls they produced.

Although biotic characteristics of an environment (i.e., the presence of heterospecific competitors) as well as the location of habitats on the landscape can influence habitat selection in adults and resulting offspring fitness, I was also interested in how abiotic characteristics can affect habitat use in other amphibians and life stages. In collaboration with David Burke of the Holden Arboretum and using a long-term manipulation of forest soil pH, I tested how forest acidification affects juvenile American toads and their interaction with the invertebrate community. While the effects of acidification on certain aspects of forest ecosystems have been well studied, less is known about the influence of soil acidification on the forest floor food web that includes amphibians and invertebrates. I evaluated survival, growth, and diet of newly metamorphosed toads placed in terrestrial enclosures in forest plots with either experimentally elevated soil pH or untreated, acidified soils. Toads tended to grow larger in elevated soil pH, although survival and diet were not affected by pH. I also found no effect of pH on the invertebrate community in the or forest floor trophic dynamics.

Importantly, the presence of temporary terrestrial mesocosms we constructed significantly reduced invertebrate abundances and overall diversity. Thus, the strong effect these structures can have on invertebrate communities should be considered when used in future studies.

Chapter 1: Female treefrog preference for breeding sites matches offspring performance in the presence of two anuran competitors

Authors: David A. Dimitrie[1], Michael F. Benard[1]

1. Department of Biology, Case Western Reserve University, Cleveland, OH

INTRODUCTION

During the breeding season, animals in heterogeneous landscapes must make

decisions about where to place their offspring (Schmidt et al 2010). Preference-

performance theory predicts that if habitat selection is adaptive, female preference for

breeding sites should maximize offspring performance (e.g., the "mother knows best"

hypothesis in insect herbivores; Garcia-Robledo and Horvitz 2012). Understanding if

these behaviors are adaptive for offspring can have important implications for linking

habitat selection to local adaptation, population dynamics, community assembly, and

biodiversity patterns (Spencer et al. 2002, Morris 2003, Binckley and Resetarits 2005).

Evidence both for and against adaptive habitat selection for offspring performance

exists in a variety of systems. Adaptive habitat selection that maximizes offspring fitness

occurs in organisms ranging from damselflies to mosquitoes and birds (Morosinotto et al.

2010, Yoshioka et al. 2012, Lambret et al. 2018). However, offspring performance does

not always match female site preference, as females may select habitats which increase

their own longevity while reducing their offspring survival (i.e., the "optimal bad

motherhood" principle; Garcia-Robledo and Horvitz 2012). In other cases, breeding

behavior may be maladaptive for offspring due to ecological traps resulting from

anthropogenic disturbances (Schlaepfer et al. 2002). Habitat selection may also be neutral

and result in no fitness differences for offspring (Silberbush and Resetarits 2017), or

breeding females may not be able to distinguish among habitats of varying quality and

thus not demonstrate any evidence of habitat selection behavior (Buxton and Sperry

2017). Competitors and predators can specifically alter the quality of a habitat for a

breeding female. For example, conspecifics and predators can affect nesting site selection

in birds by decreasing successful mating attempts and decreasing hatching success

(Morosinotto et al. 2010, Mueller et al. 2016). However, evaluating the female preference-offspring performance relationship in the presence of nonpredatory heterospecific competitors is an area of research that is lacking in studies of breeding site selection. Understanding this relationship in the context of heterospecific competitors can have broader ecological and evolutionary consequences.

Because most species of anurans deposit their eggs in aquatic habitats and provide no further parental care, they are an excellent study system to investigate female preference-offspring performance theory. Buxton and Sperry (2017) reviewed studies on biotic factors influencing female anuran breeding habitat selection, and the presence of predators and conspecifics comprised most studies. There are several well-described examples of female preference for certain types of habitats, such as avoiding breeding habitats containing certain predators (Rieger et al. 2004) and older conspecific competitors (Resetarits and Wilbur 1989). However, the relatively few studies that have tested if anurans prefer to breed in habitats based on the presence of heterospecific anurans have produced mixed results including attraction to, avoidance of, and no preference (Resetarits and Wilbur 1989, Indermaur et al. 2010, Dananay and Benard 2018). Tests of whether anuran preference for habitats is correlated with offspring performance are less common, and these studies have also largely focused on predator and conspecific effects on resulting offspring performance. For example, females avoid breeding habitats containing fish due to an increase in larval mortality in these habitats (Rieger et al 2004). Females will also avoid habitats with conspecific larvae to minimize intraspecific competition for their offspring (Pintar and Resetarits 2017).

The responses of breeding anurans to tadpoles of other species are much more variable than for predators such as fish or competitors such as older conspecifics. Many studies of responses to heterospecifics investigate avoidance of tadpoles that prey on anuran eggs; however, direct tests of female breeding site preference in the presence of nonpredatory heterospecifics are minimal and suggest they do not always lead to female avoidance (Buxton and Sperry 2017). Nonpredatory heterospecifics may negatively affect offspring performance, but they may also have positive effects on offspring performance (Mackey and Boone 2009) that can result in increased adult fitness following metamorphosis (Altwegg and Reyer 2003). Therefore, linking female breeding habitat preference to offspring performance among habitats containing heterospecific anurans can have important implications for both anuran population dynamics and broader community structure on the landscape.

We tested the effects of two anuran heterospecifics, the green frog (*Rana clamitans*) and the bullfrog (*R. catesbeiana*), on breeding habitat preference and offspring performance in the eastern grey treefrog (*Hyla versicolor*). We chose green frogs and bullfrogs because their larvae can occupy the same habitats as the gray treefrog, but differences in their traits suggest that bullfrogs may have a greater negative effect on treefrog tadpoles than green frogs and may thus more strongly influence breeding behavior. Although both species are predominantly omnivores that consume algae, diatoms, invertebrates, and even other amphibian eggs (e.g., Petranka and Kennedy 1999), Schiesari et al. (2009) suggested that larval bullfrogs may function primarily as a carnivorous predator under certain ecological conditions. Green frog larvae are also typically less active than bullfrog tadpoles (Werner 1991; but see Anholt et al. 2000), and green frog tadpoles can have no effect or even a positive effect on gray treefrog tadpoles

(Smith et al. 2004, Mackey and Boone 2009). Thus, bullfrog tadpoles could have stronger

negative competitive effects on gray treefrog tadpoles based on these differences in diet

and activity.

We conducted two experiments to test female preference and offspring

performance in the eastern gray treefrog. First, we set up a female site selection study

using aquatic habitats containing green frog tadpoles, bullfrog tadpoles, or no

heterospecific tadpoles and tested for differences in frequency of use, time to first use,

and total number of eggs deposited among the habitats. We then tested if female

preference for habitats matched the performance of gray treefrog offspring using a

tadpole competition experiment where we raised treefrog tadpoles in the presence and

absence of green frog and bullfrog tadpoles. We evaluated tadpole performance by

measuring larval survival, development rate, and growth.

METHODS

Female oviposition site selection

We conducted our study at Case Western Reserve University's Squire Valleevue

and Valley Ridge Farms in Cuyahoga County, Ohio, USA. We set up our study in two

fields separated by 0.65 km, each adjacent to a known gray treefrog breeding pond

(hereafter, "Skating Pond" and "Tamar Pond"). We set up 1,100 L aquatic mesocosms

(hereafter "pools") as available habitat for breeding treefrogs. We placed 18 pools in each

field in a 9x2 grid, so that each pool was 10 m from its closest neighboring pool. Distance

from pools to the edge of the respective breeding pond ranged from 25-65 m. We filled

all pools on 29 April 2019, stocked each with 10 g rabbit chow and 1 kg of mixed-hardwood leaves, and left each covered with lids. Pools were randomly assigned to one of three heterospecific treatments: no tadpoles, ten green frog tadpoles, or ten bullfrog tadpoles. Each treatment was replicated six times per field. Heterospecific tadpoles were collected from two ponds adjacent to the fields and added to pools between 10 May and 12 May. All tadpoles had overwintered at least one year in their original pond, and total competitor biomass per pool was similar between both treatments (green frog average 30.6 g (range 28.2-34.8 g); bullfrog average 33.6 g (range 31.8-34.7 g)). We removed the lids from all pools on 19 May and checked each pool for gray treefrogs eggs every 1-2 days. Eggs were removed, placed in a tray, photographed, and the number of eggs from each event was counted from the photo. Eggs were not returned to their pool to eliminate the effect of conspecific tadpoles. We continued checks until 12 July 2019, at which time we concluded the experiment.

We tested for effects of heterospecifics on total number of pools used at least once by females using a chi-squared test of independence on a contingency table with heterospecific treatments and number of pools used and unused at the end of the study. We also evaluated the effects of heterospecifics on the total number of nights each pool was used to assess the overall attractiveness of each treatment. In anurans with an extended breeding season, males may be active most nights of the breeding season while female activity is restricted to a subset of nights (Rieger et al. 2004). Thus, we assessed the use of pools on those nights when females were specifically present and using our fields (at least one oviposition event in at least one pool on a given night; hereafter "active female night"). Female breeding was documented in Skating Pond field and Tamar Pond field on 14 and 6 nights, respectively, of 54 sample nights. We therefore categorized those as active female nights in each field. We used a chi-squared test of independence to test the effects of heterospecifics on the number of nights a pool type was used among all active female nights in each field.

We tested for an effect of heterospecifics, field, and their interaction on time to first use of pools and total number of eggs deposited in each pool using separate generalized linear models with a negative binomial error distribution. The interaction between treatment and field was not significant ($P = 0.09$ and $P = 0.90$, respectively) and was dropped from both models. Models were fit using the *glm* function and followed with post-hoc comparisons using the *emmeans* function in the *emmeans* package to test for specific differences among the three treatments.

Larval performance

To test if these heterospecifics affect larval treefrog performance, we conducted a mesocosm competition experiment. We established 30 pools and assigned each to one of three heterospecific treatments: no tadpoles, ten green frog tadpoles, or ten bullfrog tadpoles. Each treatment was replicated 10 times. Pools were placed in five rows in a field so that each row contained six pools, with two replicates of each treatment randomly assigned to each row. Each pool was filled on 21 May 2019 and stocked with 10 g rabbit chow and 1 kg of mixed-hardwood leaves. We collected heterospecific tadpoles on 29 May from the same source ponds as those in the female site selection study. Treefrog tadpoles originated from six egg masses collected on four nights (7 June, 9 June, 15 June, and 19 June) during the habitat selection study. Between June 20 and June 23, we haphazardly added 50 hatchlings to each pool, such that each pool received 8-10 tadpoles from each of the egg masses. We then ended the experiment between 16 July and 18 July when we observed the first signs of treefrog tadpoles approaching metamorphosis (forelimbs emerged). Pools were emptied and tadpoles were euthanized in MS-222 and stored in 70% ethanol. We then quantified number of tadpoles surviving per pool and measured developmental stage, body mass, and snout-vent length (SVL) for each tadpole.

We tested if heterospecifics affected the number of tadpoles surviving using a generalized linear model. We used the *cbind* function to evaluate our response variable, survival, using the number of tadpoles located and not located per pool at the end of the study. We fit the model using the *glm* function with treatment as our explanatory variable and a quasibinomial error distribution. We used linear mixed-effect models to test whether heterospecifics affected tadpole developmental stage, body mass, and SVL. Models included heterospecific treatment as the explanatory variable and pool as a random effect nested within stratified row in the study field. Mixed-effect models were fit using the *lmer* function in the *lme4* package and followed with post-hoc comparisons using the *emmeans* function. We conducted all analyses using R software environment version 4.0.5 (R Core Team 2021).

RESULTS

Female oviposition site selection

We documented 18 unique breeding events (i.e., at least one egg mass in an individual pool) on 14 active female nights in the Skating Pond field and 8 unique breeding events on 6 active female nights in the Tamar Pond field. We found no association between the heterospecifics present and the number of pools used at least once (Supplemental Table 1). However, pools with no heterospecifics present were used on more nights than those with either heterospecific present (Table 1). Average time to first use of pools differed among the three treatments ($\chi^2 = 17.78$, $P < 0.001$; Figure 1A) and between the two fields ($\chi^2 = 5.28$ $P = 0.02$). Pools with no heterospecifics were first used on average 12 and 19 days earlier than those with green frog and bullfrog tadpoles, respectively, and pools were first used in the Skating Pond field after an average of 31 days compared to 40 days in the Tamar Pond field. We found a significant difference in total number of eggs among treatments ($\chi^2 = 25.04$, $P < 0.001$; Figure 1B) but not between fields ($\chi^2 = 0.11$, $P = 0.75$). Three times as many eggs were found in pools without heterospecifics, but there was no difference between the green frog and bullfrog treatments.

Larval performance

Tadpole survival was not affected by heterospecifics and averaged 67% across all treatments ($\chi^2 = 0.65$, $P = 0.72$, Figure 2A). However, heterospecifics had a significant effect on tadpole development ($\chi^2 = 30.10$, $P < 0.001$), body mass ($\chi^2 = 31.41$, $P < 0.001$) and SVL ($\chi^2 = 31.74$, $P < 0.001$). Tadpoles in pools with no heterospecifics were 11% more developed than those in pools with either heterospecific (Figure 2B). Tadpole body mass in pools with no heterospecifics was nearly twice that in pools with either heterospecifics (Figure 2C), and tadpole SVL was 26% larger in the absence of heterospecifics (Figure 2D). Tadpole development, body mass, and SVL did not differ between the green frog and bullfrog treatments.

DISCUSSION

If habitat selection is adaptive, female preference for breeding sites should maximize offspring performance. For example, predators and competitors can affect female habitat selection in birds, such that utilizing habitats without them provides a reproductive benefit via mechanisms such as an increase in successful mating attempts, increased hatching success, and increased offspring performance (Morosinotto et al. 2010, Mueller et al. 2016). However, female preference does not always benefit offspring, as has been demonstrated in some insects (Garcia-Robledo and Horvitz 2012). In anurans, the effects of heterospecific competitors on female breeding have received less attention than the effects of conspecifics and predators (Buxton and Sperry 2017), yet heterospecifics may influence female preference and resulting offspring performance. We found that female gray treefrogs used pools with no heterospecifics more frequently and earlier in the breeding season than pools with heterospecifics. Tadpoles raised in the presence and absence of these heterospecifics demonstrated that this female preference is adaptive, as tadpoles grew larger and developed faster in the absence of heterospecifics. Metamorphosing earlier and larger can increase post-metamorphic fitness (e.g., survival and growth; Altwegg and Reyer 2003). Our results demonstrate that female breeding site selection is an adaptive behavior for offspring performance in the presence of nonpredatory heterospecific competitors, an area of research that is lacking in habitat selection studies. These finding are important for understanding the role of breeding behavior in heterogeneous landscapes where heterospecific distributions are spatially variable, which can play a critical role in population and community dynamics (Morris 2003).

Most studies of female preference in anurans have tested the effects of predators or conspecifics on female choice, while fewer have tested the effects of nonpredatory heterospecifics (Buxton and Sperry 2017). Resetarits and Wilbur (1989) documented no avoidance of pools with bullfrog tadpoles by female Cope's gray treefrogs (*Hyla chrysoscelis)*. Interestingly, the authors used a much larger starting density of overwintered bullfrogs (100 tadpoles compared to 10 tadpoles in our study) and yet found no difference in female activity between empty pools and pools with bullfrogs. This may be due to the differences in the ecological communities presented to treefrogs in these two studies. Resetarits and Wilbur (1989) provided seven different habitat choices. These included known predators (larval salamanders or fish) or conspecific tadpoles, all of which were perceived as lower quality habitats compared to habitats with bullfrogs and thus elicited an avoidance response by female treefrogs. It is possible that habitats with predators and conspecifics were perceived as the most dangerous, and the effects of heterospecific competitors such as bullfrogs were muted even though these competitors still pose a risk to offspring. In contrast, we presented females with only three habitat choices, where pools with any heterospecifics were perceived as lower quality habitats. The number of options presented can influence the behavior of subjects in mate choice experiments, as female mating preferences are weaker when more options and extreme choices are presented (Dougherty 2020). Treefrogs appear able to exhibit similar habitat selection responses depending on the choices of habitat quality available. Presenting more options that include extreme choices (i.e., predators and conspecifics) may mute the response to less extreme choices (i.e., heterospecifics).

A critical component to any test of the preference-performance theory is explicitly measuring offspring performance in the habitats being evaluated for female preference. We hypothesized that bullfrogs would more strongly affect female preference and offspring performance than green frogs based on previous studies on diet and activity levels that indicate bullfrogs may be a stronger competitor (Werner 1991, Schiesari et al.

2009). However, females only distinguished between pools with and without

heterospecifics and not between pools with bullfrogs and green frogs. Our tadpole

performance results contrast with other studies that have tested the effects of green frog

larvae on gray treefrogs but are consistent with the effects of bullfrog larvae on other

amphibian species. Smith et al. (2004) reported no effect of green frog tadpoles on gray

treefrog tadpole growth, but these green frog tadpoles were not overwintered tadpoles

and were of smaller size than the treefrog tadpoles at the start of their study. Our study

used larger, overwintered green frog tadpoles which may explain the differences we

observed. However, Mackey and Boone (2009) documented a positive effect of

overwintered green frog tadpoles on gray treefrog development. In contrast, overwintered

bullfrog tadpoles negatively affect size at metamorphosis in several amphibian species

(Boone et al. 2004), consistent with our findings. The differences between our tadpole

results and previous work may be due to differences in experimental venues, but,

importantly, our results align with female preferences for breeding sites and thus support

the preference-performance theory.

It is possible that in natural wetlands green frog and bullfrog tadpoles act as

predators on treefrog eggs or tadpoles, and thus it is predation, not competition, that

influenced female preference-offspring performance. Larvae of both species will

consume amphibian eggs (e.g., Petranka and Kennedy 1999), and Schiesari et al. (2009)

suggested that larval bullfrog may also function primarily as a carnivorous predator under

certain ecological conditions. Overwintered bullfrogs will consume both southern leopard

frog and American toad tadpoles (Boone et al. 2004). However, our results provide

evidence that neither species consume gray treefrog larvae, as survival was similar between those tadpoles raised in the presence or absence of either heterospecific. These heterospecifics would therefore be best categorized as competitors, and our study helps address the gap in the understanding of the effects of nonpredatory heterospecifics on breeding site selection in females (Buxton and Sperry 2017).

Female treefrogs in our study used pools without heterospecifics present more frequently and earlier in the breeding season than those with heterospecifics. Time to first use can have important implications, as earlier breeding provides offspring with an advantage over later breeders through reduced competition and earlier access to resources (Wilbur and Alford 1985). Although we removed all eggs to keep conspecific densities at zero, access to resources among our pools might still be highest for those females breeding earlier in the season, especially in those pools with heterospecific tadpoles that are utilizing the same resources. It is also possible that heterospecifics in the aquatic environment can increase periphyton biomass and thus overall resource availability through their grazing activity, as demonstrated by the foothill yellow-legged frog (*Rana boylii*) in a lotic system (Kupferberg 1997). Thus, female treefrog preference for pools with green frog and bullfrog tadpoles in our study may have increased due to an increase in the availability of resources for offspring later in the season. Later in the breeding season, it may also be less important for females to assess multiple sites with different competitors present and more important to instead select any habitat to minimize the costs associated with sampling (e.g., increased chances of predation), as has been suggested for birds (Ahlering et al. 2010).

We experimentally isolated one environmental factor (i.e., heterospecific tadpoles) in our pools, but natural wetlands will have a more complex array of factors to affect female choice. For example, our design precluded the use of pools by adults of our focal heterospecifics. Adult green frogs and bullfrogs will consume treefrogs, so their presence could further influence female preference and offspring performance. Heterospecific tadpoles can also interact as part of more complex ecological communities (e.g., with fish or invertebrates), and these communities could in turn affect habitat selection and offspring performance. Finally, male behavior can play a role in determining where mating and egg deposition occur (Chapter 2, this book). Understanding these dynamics between breeding site preference and offspring performance is important for linking fundamental questions in habitat selection to population dynamics, community assembly, and broader biodiversity patterns.

TABLES

Table 1. Contingency table and resulting chi-squared tests for number of nights with at
least one oviposition event and number of nights with no oviposition events among
heterospecific treatments in the two fields on active female nights (i.e., any night when
oviposition occurred in at least one pool in the field; n=14 nights in Skating Pond field
and n=6 nights in Tamar Pond field).

Study field	Heterospecifics present	Number of nights used	Number of nights unused
Skating Pond	None	8	6
$X^2 = 5.60$, df =2, $P = 0.06$	Green Frog	5	9
	Bullfrog	2	12
Tamar Pond	None	6	0
$X^2 = 14.49$, df =2, $P < 0.001$	Green Frog	0	6
	Bullfrog	1	5

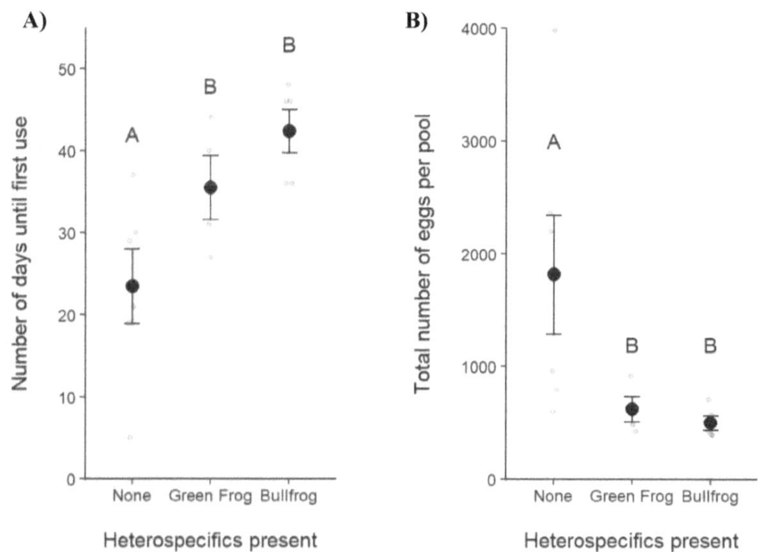

Figure 1. Effect of heterospecifics on female oviposition site selection in eastern gray treefrogs among those pools with no heterospecifics, ten green frog tadpoles, or ten bullfrog tadpoles in the two study fields. A) Number of days until first use (first oviposition event). B) Total number of treefrog eggs collected over the entire study among those pools used at least once. Data are mean ± S.E.M. for each heterospecific treatment. Individual gray points represent data for each individual pool used by females. Data with the same uppercase letter in could not be distinguished at $P < 0.05$.

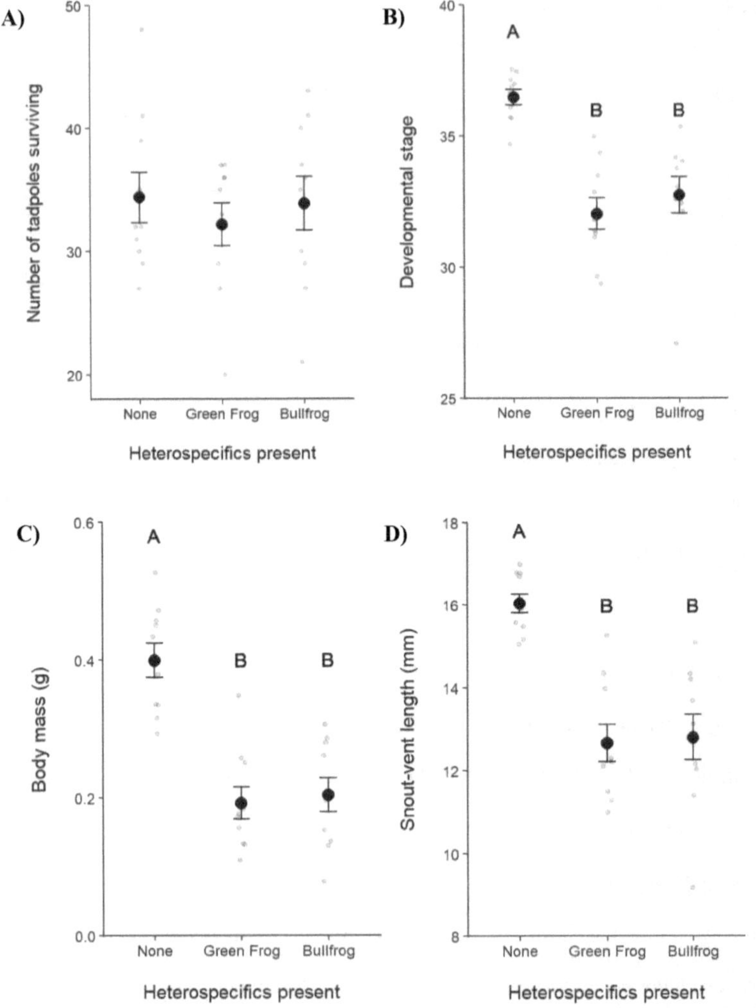

Figure 2. Eastern gray treefrog tadpole survival (A), developmental stage (B; Gosner 1960), body mass (C) and snout-vent length (D) when in the presence of no heterospecifics, ten green frog tadpoles, or ten bullfrog tadpoles. Data are mean ± S.E.M.

for each heterospecific treatment. Individual gray points represent mean values for each individual pool in each heterospecific treatment (n=10 per treatment). Data with the same uppercase letter in (B), (C), and (D) could not be distinguished at $P < 0.05$. There was no effect of heterospecific treatment on number of tadpoles surviving (A).

SUPPLEMENTAL TABLES

Supplemental Table 1. Contingency table and resulting chi-squared tests for number of ponds used and number of ponds unused throughout the entire study period among heterospecific treatments in the two study fields.

Study field	Heterospecifics present	Number of ponds used	Number of ponds unused
Skating Pond	None	4	2
$X^2 = 0.468, P = 0.79$	Green Frog	4	2
	Bullfrog	3	3
Tamar Pond	None	2	4
$X^2 = 2.57, P = 0.28$	Green Frog	0	6
	Bullfrog	2	4

Chapter 2: Effects of heterospecific anurans on calling site choice and advertisement calls of male treefrogs

Authors: David A. Dimitrie[1], Michael F. Benard[1]

1. Department of Biology, Case Western Reserve University, Cleveland, OH

INTRODUCTION

In heterogeneous environments, animals face decisions about how to use habitats of varying quality (Dall et al. 2005. Schmidt et al. 2010). An important aspect of habitat selection is that individuals assess habitat quality correctly in order to select a habitat that maximizes their fitness (Morris 2003). When multiple habitats are present to choose from, costs and constraints must be accounted for by an animal when determining if it should use a certain habitat over another (Acker et al. 2017), and this can be especially true for selection among nesting and breeding habitats (Lima 2009, Buxton and Sperry 2017). Within a species, there may be differences among individuals in habitat use due to either differences in choice or differences in ability to use a habitat. For example, males and females may use similar or different cues to assess habitat quality when selecting a breeding location (Resetarits and Wilbur 1991, Eterovick and Ferreira 2008). Whether male and female responses to habitat quality are similar and coordinated with one another or different and in conflict with one another is important for understanding why individuals use the habitats they do during the breeding season. In addition to differences between males and females, differences in the quality of individuals may be important in determining which habitats individuals use (Bowler and Benton 2005). Phenotypically inferior individuals may select or be forced to use inferior habitats to avoid competition with superior individuals (i.e., the ideal despotic distribution; Fretwell 1972).

Males should potentially align with female preferences when selecting a breeding habitat because males that select habitats that are not preferred by females may not have the opportunity to mate in a given breeding season (Eterovick and Fernandes 2002). For males using signals to attract a mate, once at the best habitats it may be advantageous to produce the best advertisement that is necessary to attract a female at that location. Aside from attracting mates, there are other adaptive reasons why males should care about what type of habitat they advertise from. For example, female anurans preferentially use habitats without certain competitors and predators present (Resetarits and Wilbur 1989, Smith and Harmon 2019), and breeding in these habitats increases offspring performance compared to offspring developing in the presence of these same competitors and predators (Rieger et al. 2004, Chapter 1, this book). Therefore, males should also use these same breeding habitats to increase both their potential of obtaining a mate and their resulting offspring success in that habitat. However, males and females may not use the same cues when selecting among potential breeding habitats, and potential sites may be of different perceived quality to males and females (Murphy 2003).

In addition to habitat quality, individual quality of males and females (e.g., developmental stage, body size/body condition, competitive ability) can be important in determining which habitats individuals use (Bowler and Benton 2005). This may be especially important during the breeding season. Purchase and Hutchings (2008) found that larger trout prevent other, smaller trout from occupying the most preferred spawning habitats. For species that signal to attract females during the breeding season, signal quality of a male can be a specific indicator to a female of the quality of that male as a parent for her offspring (Kroodsma and Byers 1991, Welch et al. 1998). It may therefore be expected that male which produce inferior signals would be forced to use low-quality habitats if higher quality males are able to use available high-quality habitats. However, the benefits conferred by males producing signals of certain quality can be context dependent (Welch 2003). Although much is known about how females select among males producing different signals in a variety of taxa and how the signal of a male indicates his quality as a potential mate, the use of habitats of varying quality by males producing signals of different quality is an understudied area of habitat selection research.

Because most species of anurans deposit their eggs in aquatic habitats and provide no further parental care, they are an excellent study system to investigate the effects of habitat quality on breeding site selection. Lack of parental care means that larvae are left at the sites selected by their parents, and thus they cannot leave if habitat conditions are unfavorable. Therefore, there should be strong selective pressures on reproducing individuals to discriminate between high- and low-quality breeding sites. Although much work has been done on female oviposition preferences among habitats of different quality (Rieger et al. 2004, Rudolf and Rödel 2005, Smith and Harmon 2019), few studies have investigated the potential for male and female behavior to be either coupled or to conflict with one another (Resetarits and Wilbur 1991, Eterovick and Ferreira 2008). This is important in frog species with an extended breeding season because males and females may be reproductively active on multiple nights and may therefore use multiple habitats over the season. In extended breeders where male frogs produce advertisement calls to attract females, selection should also favor males which call from the preferred breeding sites of females to maximize their likelihood of obtaining a mate. Because advertisement calls are energetically costly (Taigen and Wells 1985), it is advantageous for male frogs to identify the preferred habitats of females and to only call from them on nights when receptive females are present. However, males call from breeding sites on most nights during the breeding season, even when females are not present (Fellers 1979). Therefore, in terms of active breeding nights, male behavior is not coupled with females. Males should therefore identify and establish a calling territory at the preferred location for female oviposition so that they can call from these habitats when females are in fact present.

The eastern gray treefrog (*Hyla versicolor*) is a well-studied system in both habitat selection (Resetarits 1996) and anuran communication (Gerhardt and Huber 2002). Females preferentially select habitats lacking certain offspring predators (Smith and Harmon 2019) and older conspecifics (Resetarits and Wilbur 1989) as well as some heterospecific anurans that can act as competitors with treefrog offspring (Chapter 1, this book). Males show similar preferences in habitats containing some of the same potential offspring predators and competitors, but these preferences are not always the same. For example, both males and females will avoid breeding habitats with fish or older conspecific tadpoles, but only females will avoid the predatory larvae of *Ambystoma* salamanders (Resetarits and Wilbur 1991). In general, the conditions under which male and female decisions are coupled or in conflict remain ambiguous. Additionally, male gray treefrogs produce distinct advertisement calls consisting of repeated pulses that are used to attract females during the breeding season. In choice experiments, females prefer longer calls to shorter calls as well as males spending more time actively calling (Schwartz et al. 2001). Females mating with males giving longer calls also produce offspring that perform better in their developmental habitat (Welch et al. 1998). However, these long callers do not benefit offspring in all environments. Offspring produced by long callers perform better than those of short callers at a low larval density, but these benefits do not occur for offspring raised in high larval density conditions (Welch 2003). Female preference for males producing calls with certain characteristics in certain habitats makes the eastern gray treefrog an ideal study organism for investigating how males advertise to attract a female from breeding habitats of different quality.

In this study we tested if male and female use of habitat cues are coupled or in conflict with each other and if male eastern gray treefrogs of lower quality select and advertise from habitats of lower quality. We conducted an experiment to test male preference for breeding habitats with and without heterospecifics present and to compare these preferences to female breeding preference and offspring performance. We tested the effects of two larval heterospecifics, the green frog (*Rana clamitans*) and the bullfrog (*R. catesbeiana*), on male breeding habitat preference in the eastern grey treefrog (*Hyla versicolor*). We have documented that both these heterospecifics act as competitors of gray treefrog larvae and elicit a negative effect on female gray treefrog breeding site selection and offspring performance in our study system (Chapter 1, this book). By demonstrating this female preference and resulting offspring performance in habitats without either of these competitors, we were then able to examine how males use these same available habitats when selecting advertisement locations to attract females. This allowed us to test if male and female use of larval competitor cues among available breeding habitats are coupled or in conflict with one another. We set up a male site selection study using aquatic habitats containing green frog tadpoles, bullfrog tadpoles, or no heterospecific anurans and tested for differences in frequency of use and time to first use by breeding males. We also tested for differences in the quality of these males as potential offspring sires by comparing their body size and advertisement call characteristics among the three aquatic habitat types.

METHODS

Male breeding site selection

We conducted our study at Case Western Reserve University's Squire Valleevue and Valley Ridge Farms in Cuyahoga County, Ohio, USA. We set up our study in two fields separated by 0.65 km, each adjacent to a known gray treefrog breeding pond (hereafter, "Skating Pond" and "Tamar Pond"). We set up 1,100 L aquatic mesocosms (hereafter "pools") as available habitat for breeding treefrogs. We placed 18 pools in each field in a 9x2 grid, so that each pool was 10 m from its closest neighboring pool. Distance from pools to the edge of the respective breeding pond ranged from 25-65 m. We filled all pools on 29 April 2019, stocked each with 10 g rabbit chow and 1 kg of mixed-hardwood leaves, and left each covered with lids. Pools were randomly assigned to one of three heterospecific treatments: no tadpoles, ten green frog tadpoles, or ten bullfrog tadpoles. Each treatment was replicated six times per field. Heterospecific tadpoles were collected from two ponds adjacent to the fields and added to pools between 10 May and 12 May. All tadpoles had overwintered at least one year in their original pond, and total competitor biomass per pool was similar between both treatments (green frog average 30.6 g (range 28.2-34.8 g); bullfrog average 33.6 g (range 31.8-34.7 g)). We removed the lids from all pools on 19 May and checked each pool for male gray treefrogs on most nights throughout the breeding season (n=43 nights). For each located male, we documented the pool being used, took a photo of the frog dorsum to use in identification of each individual frog throughout the breeding season, and measured body mass (on most nights; e.g., excluding nights when a male was located in amplexus with a female to avoid disrupting the mating). We continued checks until 11 July 2019, at which time we concluded the experiment.

Male advertisement call characteristics

We recorded advertisement calls from males at each pool on most nights when males were calling (n=27 nights with at least one recording). We recorded males using a Marantz PMDF 661 field recorder with a Sennheiser ME66/K6P shotgun microphone. We set the microphone on a tripod boom microphone stand approximately 0.5 m from the focal male, began recording, and left the recording area to eliminate any influence on the male's advertisement call. We then allowed the recording to proceed for at least 3 minutes when possible. We took dorsum photos of each frog and collected body mass measurements following the recording to eliminate the effects of animal handling on call characteristics. We also used an analog pocket thermometer (W. W. Grainger, Inc.) to measure the ambient temperature adjacent to the calling frog immediately following the recording. For each recording, we then calculated average call duration, average number of pulses per call, and overall call effort (proportion of time spent calling) on a 1-minute portion of each recording (generally during the middle of each recording when frog calling behavior was stable). Call characters were measured using Raven Pro 1.6 software (Center for Conservation Bioacoustics, Cornell Lab of Ornithology).

Statistical analyses

We tested for effects of heterospecifics on the total number of pools used at least once by males using a chi-squared test of independence on a contingency table with heterospecific treatments and number of pools used and unused at the end of the study in each field. We tested for an effect of heterospecifics, field, and their interaction on the total number of nights that each pool was used and on the number of nights until a pool was first used using separate generalized linear models with a negative binomial error distribution. The interaction between treatment and field was not significant ($P = 0.83$ and $P = 0.97$, respectively) and was dropped from both models. Models were fit using the *glm* function and followed with post-hoc comparisons using the *emmeans* function in the *emmeans* package to test for specific differences among the three treatments.

To test if there was a difference in size between frogs using the different pool types, we used a linear mixed-effect model with frog body mass as our response variable, heterospecific treatment as a fixed effect, and a random intercept of field and individual frog ID nested within field (1|field/frog ID). We included frog ID as a random effect because some individual males were measured on multiple nights (n=13 males in Skating Pond field, range=2-9 nights per male; n=10 males in Tamar Pond field, range=2-7 nights per male), and we nested frog ID within field to account for differences between the two fields. Frogs did not move between fields in our study (based on examination of dorsum photos). We used a series of linear mixed-effect models to analyze whether heterospecific treatment affected male advertisement call characters. Specifically, we tested whether treatment affected average call duration, average number of pulses per call, and call effort. For each call character, we included the call character as the response variable, and heterospecific treatment, frog body mass, and calling temperature as fixed effects in each model. We included both the individual fixed effects of these variables as well as all potential interactions among them in our models. The models also included a random intercept of field and individual frog ID nested within field (1|field/frog ID) as described above for our frog body mass model. We included both frog mass and calling temperature in our models because both can affect advertisement call characteristics in treefrogs (e.g., Gayou 1984, Garcia et al. 2019). Models were fit using the *lmer* function in the *lme4* package in R. If body mass or call temperature were a significant predictor of a call character, we performed post-hoc linear regressions to quantify their relationship with the appropriate call character. We also excluded any calls from males utilizing the same pool as at least one other male on any night because the presence of a competing male at the same location can affect call characters of focal males (Wells and Taigen 1986, Gerhardt 1991). We wanted to assess all males using the pools within the same social context (i.e., no adjacent competing males at their pools) and therefore only used recordings from pools with one male on a given night. In total, we included 82 unique calls in our analyses. We conducted all analyses using R software environment version 4.0.5 (R Core Team 2021).

RESULTS

Male breeding site selection

We documented 20 unique individual males using the Skating Pond field on a total of 35 nights and 20 unique individual males using the Tamar Pond field on a total of 22 nights. We found no association between the heterospecific tadpoles present and the number of pools used at least once (Skating Pond $X^2 = 0.55$, df $= 2$, $P = 0.76$, Tamar Pond $X^2 = 0.47$, df $= 2$, $P = 0.78$; Table 1). Pool use was limited early in the season in both fields until approximately 9 June, when we began observing at least one male in each field on most nights (Figure 1). There were no differences in the average number of nights each pool was used among the heterospecific treatments ($X^2 = 0.65$, df $= 2$, $P = 0.72$; Figure 2A). There was also no difference in the average number of nights pools were used between the two fields ($X^2 = 1.02$, df $= 1$, $P = 0.31$). Pools were each used on average 4 nights throughout the season. Average time to first use of pools did not differ among the three treatments ($X^2 = 4.42$, df $= 2$, $P = 0.11$; Figure 2B) or between the two fields ($X^2 = 0.32$, df $= 1$, $P = 0.57$). Pools were first used after an average of 25 nights.

Male advertisement call characteristics

We measured body mass and analyzed call characters from 14 unique individual males in Skating Pond field and 15 unique individual males in Tamar Pond field. Frog mass did not differ among calling males using the three pool treatments ($X^2 = 0.43$, df $= 2$, $P = 0.81$; Figure 3). Frog body mass averaged 7.86 g (range $= 4.92$-11.07 g; S.D. $= 1.80$). For those males calling from pools on multiple nights (18 out of 29 frogs), 8 moved between two treatment pool types during the season (n $= 5$ males in Skating Pond field and n$=3$ males in Tamar Pond field; Figure 4). The remaining 11 frogs called from the same pool type each night they were present, regardless of how many nights they were present in the fields. Heterospecific treatment did not affect any call characters for the 29 recorded frogs (Table 2; Figure 5). Call temperature significantly affected both average call duration ($\chi^2 = 48.90$, $P < 0.001$) and average number of pulses per call ($\chi^2 = 9.76$, $P < 0.01$). Average call duration was negatively associated with call temperature ($F_{1,80} = 43.03$, $P < 0.001$; $R^2 = 0.34$; Figure 5A), as was average number of pulses per call ($F_{1,80} = 9.05$, $P < 0.01$; $R^2 = 0.09$; Figure 5B). There were 10 nights in each field when males were recorded from multiple pool types on the same night. However, total number of recordings each of those nights were low (maximum of 5 recordings per field per night; mean of 3 recordings per night in both the Skating Pond and Tamar Pond fields), limiting our ability to specifically test for differences in call characters among the treatments on those nights. Visual examination of call characters among the three treatments on those specific nights did not suggest any overall differences in the use of these pool types by males with different call qualities (Supplemental Figures 1-6). We documented 7 competitive interactions (≥ 2 males calling from the same pool) on 6 nights. Although these calls were not included in our analyses, we documented 3, 1, and 3 of these interactions in the treatments containing no heterospecifics, green frog tadpoles, and bullfrog tadpoles, respectively. Thus, the number of competitive interactions that occurred was apparently not affected by the presence or absence of heterospecifics in pools.

To confirm that we had the statistical power to detect a biologically meaningful difference in call characters between males using the different habitats, we conducted a post-hoc power analysis using average number of pulses per call. Using previous data demonstrating that females distinguish among calling males with equal calling effort when their average pulse rates differ by 12 pulses per call (Klump and Gerhardt 1987) and using a standard deviation of 3.1 on a mean of 14.1 pulses per call (Gerhardt et al. 1996), 4 males per treatment would be necessary to detect a difference in our study among the three treatments at 80% power (using a one-way ANOVA in our study design to only test differences in average number of pulses per call among our three treatments). Adjusting for the fact that females can distinguish among calling males with unequal call effort when their average pulse rates differ by only 6 pulses per call (Klump and Gerhardt 1987), a sample size of 12 males per treatment would be needed to detect a difference at 80% power. Thus, we had a sufficient sample size to detect biologically meaningful differences in call characters between treatments.

DISCUSSION

Habitat selection among breeding sites of varying quality is an important behavior that can affect both an individual's fitness and that of its offspring (Morris 2003, Schmidt et al. 2010). For males that select among habitats to display to attract a female, selecting habitats that are preferred by females and producing the best necessary display at that habitat would maximize his ability to obtain a mate. Females, in contrast, need to select among the best available mates as well as consider the best habitat for their offspring. Male eastern gray treefrog habitat selection in our study did not match female habitat preference for egg oviposition. We presented males with breeding pools containing either of two heterospecific anurans that act as competitors for treefrog tadpoles or pools with no competitors present. We previously demonstrated that females use pools without these competitors present more for egg oviposition and that this behavior benefits offspring performance (Chapter 1, this book). However, males used these pools with offspring competitors present to advertise to females the same amount as those pools without competitors. Thus, male and female assessment and use of these competitor cues when selecting a breeding habitat are not coupled. Males may possibly not assess or be able to detect differences in habitat quality cues the way females do. Previous studies have documented that male and female frogs may use similar or different cues to assess habitat quality during the breeding season. Eterovick and Ferreira (2008) documented that male and female *Scinax* treefrogs use similar habitat cues and thus have coupled

responses in breeding habitat selection in stream habitats. These habitat cues included

both abiotic seasonal conditions (dry versus rainy season) and microhabitat locations

(ground versus substrates greater than 70 cm high). In contrast, male and female *Hyla*

chrysoscelis respond independently to breeding habitat cues when presented with habitats

containing conspecifics, heterospecific competitors, and predators (Resetarits and Wilbur

1991). A similar lack of complete correspondence between male and female selection of

breeding habitats in the presence of and absence of predators and conspecifics occurs in

the neotropical frog *Edalorhina perezi* (Murphy 2003). In the presence and absence of

predators and competitors, these previous studies demonstrate that a lack of

correspondence between male and female use of habitat quality cues may be more

common than not, and our study supports these previous findings as well.

Although selection of a breeding site is an initial key component of breeding

behavior in males, males then produce advertisements to attract females at the selected

breeding site. Higher quality male parents may be able to advertise from preferred female

habitats more than lower quality males. Differences in the quality of individuals may be

important in determining which habitats individuals use (Bowler and Benton 2005).

Inferior individuals may select or be forced to use inferior habitats to avoid competition

with males of higher quality, as has been demonstrated in organisms ranging from fish to

mammals (Kropowski 1993, Purchase and Hutchings 2008, Davidian et al 2016).

Therefore, even though there may be no difference in the absolute use of different pool

types by males in our study, that may be because the preferred pools are being used by

males producing the most attractive calls. Although the three treatment pools were used an equal number of nights by advertising male treefrogs, we hypothesized that the call characters of males using the preferred habitat of female treefrogs (pools with no competitors) would produce more attractive calls than the males calling from the less preferred habitats of the females (pools with competitors). Females prefer longer calls to shorter calls (Schwartz et al. 2001), and this results in offspring that perform better in their developmental habitat (Welch et al. 1998). However, we found that males do not exhibit a difference in call quality when advertising from habitats with and without green frog and bullfrog tadpoles. Males using the three pool types produced calls of the same length and of the same number of pulses. The males also called for the same amount of effort regardless of what pool type they were calling from. Importantly, the genetic benefits conferred by long callers to a female are not always beneficial to offspring in all environments. For example, Welch (2003) demonstrated that offspring produced by long-calling males performed better than those of short-calling males when raised at a low larval density, but these performance benefits did not occur for offspring raised in high larval density conditions. Thus, the benefits of mating with a male producing a long call are context dependent. It is possible that there is no offspring benefit for females mating with males producing longer calls in the presence or absence of the green frog or bullfrog tadpoles in our pools.

Because male competition (more than one male calling from the same pool on a given night) was limited in our study and our pools were separated by 10 meters, males calling from our tanks would be best described as isolated males. Thus, they all were assessed in the same social environment (no immediate competing males), and the heterospecific communities of the pools they were calling from were the only difference among the available habitats. The social environment surrounding an advertising male affects his call characters, where the presence of nearby males (< 4 meters from a focal male) leads to a decrease in the number of calls produced but an increase in call duration and number of pulses per call for a focal male (Wells and Taigen 1986, Gerhardt 1991). Signaling plasticity in varying social environments also occurs in other taxa including invertebrates and birds (Brumm and Slater 2006, Anichini et al. 2019). Therefore, the lack of a difference in call characters among males using the different pools in our study may not be unexpected for males calling in isolation. It may not be possible to detect a difference in call quality until males are in a more competitive social context when advertising to a female. Our results instead can be viewed as a general baseline for how isolated males call from habitats of different quality. Additional research should focus on moving males from these isolated social environments to a competitive interaction involving one or more additional calling males, and I address aspects of this question in Chapter 3. The call characters in a more competitive social environment could then be compared to these baseline call characters of isolated males at these habitats with heterospecifics present or absent. This would allow a better assessment of if males using habitats of different quality produce calls of different quality once in a competitive social context.

Some male treefrogs displayed a plastic response to calling site selection, moving between multiple pools and pool types during the season. Similar plasticity in call site selection has been documented in *H. versicolor* in response to variation in ambient temperatures and the presence of predators (Höbel and Barta 2014). These individuals may be adopting the win-stay/lose-switch habitat selection strategy (Campomizzi et al.

2012), where successfully establishing a calling location and attracting a female result in a male staying at the same pool over multiple nights, while an unsuccessful male may move among multiple pools to increase his mating success. However, based on our study design we cannot say with any certainty which males were successful in obtaining a mate and which were not. The win-stay/lose-switch strategy may also explain why some individuals were only found using the pools in the fields one night before not being located again. These individuals may have assessed and used a given pool on a single night without success before moving to the main breeding pond on subsequent nights.

Our study design did not specifically allow us to test if males mating with the females that ultimately deposited eggs in our study pools were producing more attractive calls than unsuccessful males. This type of choice test in the field would better explain what the steps are for both males and females when choosing among multiple breeding sites. Females may be assessing habitats for the best offspring conditions and then mating with a suitable male at the best available habitat without comparing among multiple males at other potential habitats. Males, in contrast, may be selecting among available habitats regardless of the larval conditions present. Instead, they may be utilizing a spacing technique to avoid competitors at their advertising location. Treefrogs shorten their calls in the absence of competing males, which is less energetically costly than longer calls (Wells and Taigen 1986, Neelon and Höbel 2019). Males may then be able to maintain calling longer each night and throughout the breeding season at these isolated locations, even though females are preferentially attracted to males giving longer calls. Wells and Taigen (1986) hypothesized that that production of long calls leads to a more rapid depletion of glycogen in the muscles that are used during calling. Therefore, the production of short calls by males in the absence of other calling males reduces muscle glycogen depletion. This would increase overall calling endurance for these males. In natural settings, the number of nights spent advertising is a better predictor of male mating success than advertisement call characters or body condition in many species of frogs, including *H. versicolor* (Gerhardt et al. 1987; Sullivan and Henshaw 1992, Bertram et al. 1996). Therefore, males calling from our pools with no interacting competitors could benefit from producing the least energetically costly call to maintain a territory at a pool over multiple nights.

Selection may favor males calling from breeding sites that are coupled with female preference and resulting offspring performance. Our previous work demonstrates that more eggs are deposited in those pools without heterospecifics present, and depositing offspring in habitats without these heterospecifics present is an adaptive behavior that improves offspring performance. However, males in our study did not use habitats with and without heterospecifics differently. Male and female use of cues, at least heterospecific competitors, does not appear to be exclusively coupled. Resetarits and Wilbur (1991) documented similar differences in a closely related species, *H. chrysoscelis*. Furthermore, males calling from our pools produced calls of similar perceived quality for females, regardless of if it was at a pool with larval competitors present or not. We hypothesize that our results may be because males do not (or cannot) differentiate between the presence or absence of heterospecifics in the larval environment, or that avoiding competition by not sharing a pool with another advertising male is more important than offspring habitat quality. Whether or not male and female responses to habitat quality are similar and coordinated or different and in conflict with one another is important for our understanding of habitat selection behavior. These behaviors between males and females can then have important implications for both population dynamics and broader community structure on the landscape.

TABLES

Table 1. Contingency table and resulting chi-squared tests for number of pools used and number of pools unused throughout the entire study period among heterospecific treatments in the two study fields (n=6 pools per treatment in each study field).

Study field	Heterospecifics present	Number of pools used	Number of pools unused
Skating Pond	None	4	2
$X^2 = 0.55$, df = 2, $P = 0.76$	Green Frog	4	2
	Bullfrog	5	1
Tamar Pond	None	3	3
$X^2 = 0.47$, df = 2, $P = 0.78$	Green Frog	4	2
	Bullfrog	4	2

Table 2. Linear mixed-effects model results for the effects of heterospecific treatment, frog mass, and calling temperature on the call characters of call duration, number of pulses per call, and call effort recorded from male gray treefrogs using pools among the three heterospecific treatments.

Call character		χ^2	P
Call duration	Treatment	0.50	0.78
	Frog mass	3.45	0.65
	Call temperature	**48.90**	**<0.001**
	Treatment * Frog mass	3.27	0.20
	Treatment * Call temperature	3.18	0.20
	Call temperature * Frog mass	1.38	0.24
	Treatment * Frog mass * Call temperature	3.80	0.15
Number of pulses/call	Treatment	1.05	0.59
	Frog mass	1.53	0.22
	Call temperature	**9.76**	**<0.01**
	Treatment * Frog mass	4.15	0.13
	Treatment * Call temperature	0.45	0.80
	Call temperature * Frog mass	0.13	0.72
	Treatment * Frog mass * Call temperature	0.70	0.70
Call effort	Treatment	0.60	0.74
	Frog mass	0.45	0.50
	Call temperature	0.09	0.76
	Treatment * Frog mass	0.20	0.20
	Treatment * Call temperature	0.45	0.80
	Call temperature * Frog mass	0.79	0.37
	Treatment * Frog mass * Call temperature	0.50	0.78

FIGURES

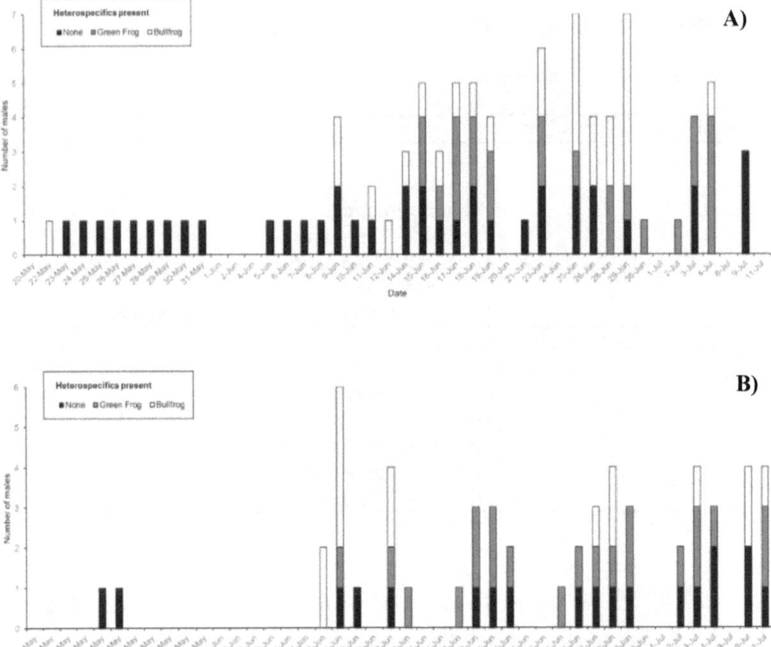

Figure 1. Number of males observed on each night within each heterospecific treatment over the entire study period in A) Skating Pond field and B) Tamar Pond field.

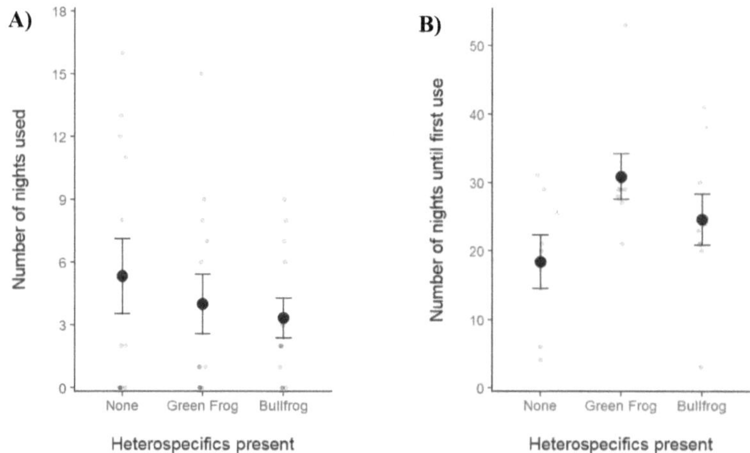

Figure 2. Effect of heterospecifics on male eastern gray treefrog breeding behavior among those pools with no heterospecifics, ten green frog tadpoles, or ten bullfrog tadpoles in the two study fields combined. A) Total number of nights each pool was used over the entire study period. B) Number of nights until first use (first male located using a pool) among those pools used at least once during the study period. Data are mean ± S.E.M per heterospecific treatment. Individual points represent data for each individual pool within each heterospecific treatment.

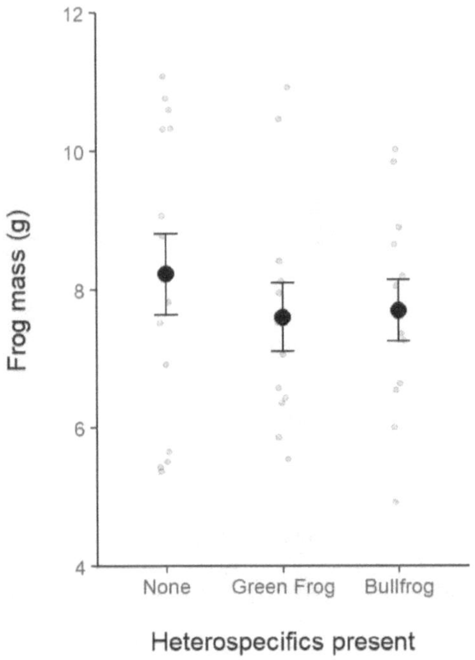

Figure 3. Average body mass among frogs using those pools with no heterospecifics, ten green frog tadpoles, or ten bullfrog tadpoles. Data are mean ± S.E.M per heterospecific treatment. Individual points represent the body mass of each frog recorded. For those males recorded on multiple nights within the same treatment, individual points represent mean frog mass for all nights he was recorded. For those males recorded in multiple treatments throughout the breeding season, mean mass in each treatment type used is represented by an individual point in each of the respective treatments used.

A)

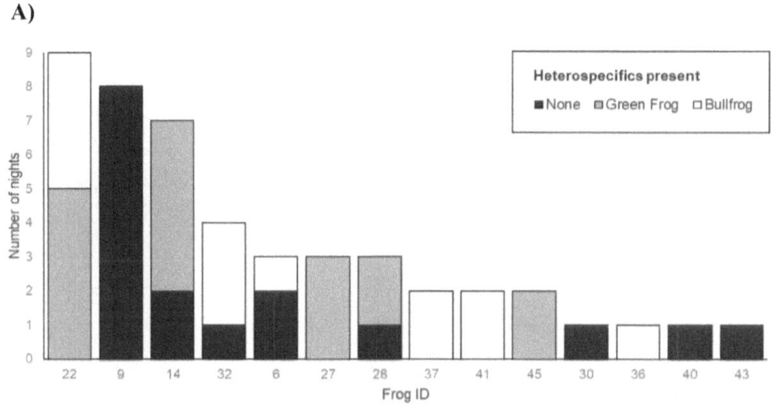

B)

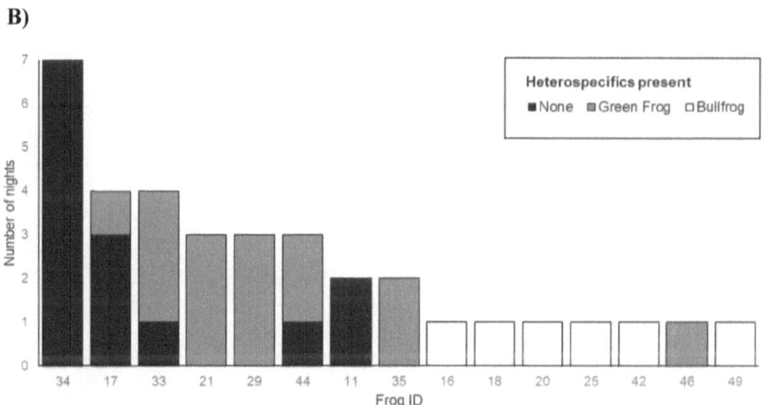

Figure 4. Number of nights each individual frog used pool types within each heterospecific treatment over the entire study period in A) Skating Pond field and B) Tamar Pond field. Frog ID represents a unique number given to each frog identified during the study.

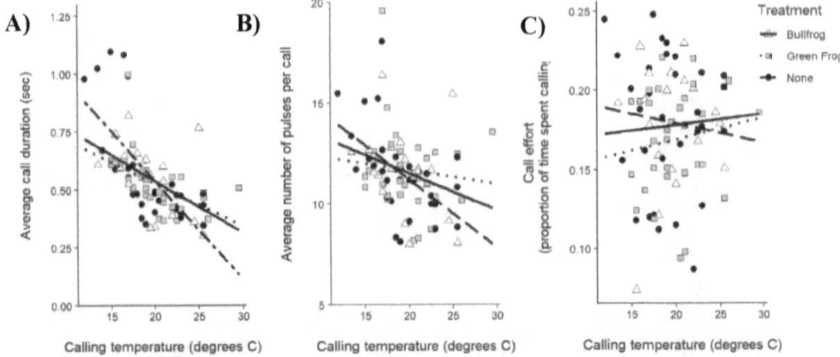

Figure 5. Call characters of male eastern gray treefrogs among those pools with no heterospecifics (filled circles), ten green frog tadpoles (gray squares), or ten bullfrog tadpoles (open triangles) in the two study fields combined. A) Average call duration, B) average number of pulses per call, and C) call effort (proportion of time spent calling). In each figure, individual points represent the average call character from each individual male recorded at a pool on each night. Lines represent the line of best fit for frog calling temperature using a liner model for each treatment for each of the three call characters.

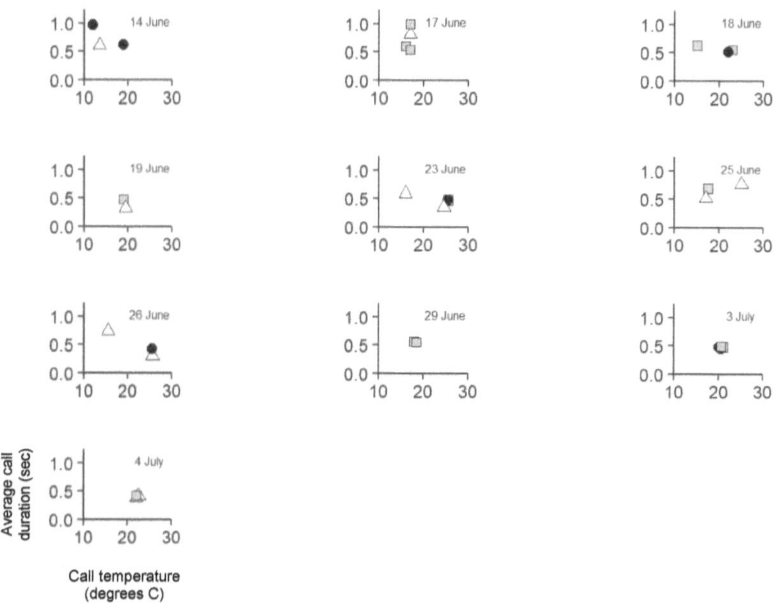

Supplemental Figure 1. Comparison of average call duration and call temperature of male eastern gray treefrogs on nights during the study when more than one male was recorded using a pool in the Skating Pond field. Males were recorded among those pools with no heterospecifics (filled circles), ten green frog tadpoles (gray squares), or ten bullfrog tadpoles (open triangles). Individual points represent the average call duration for each recorded male using a pool on the date listed.

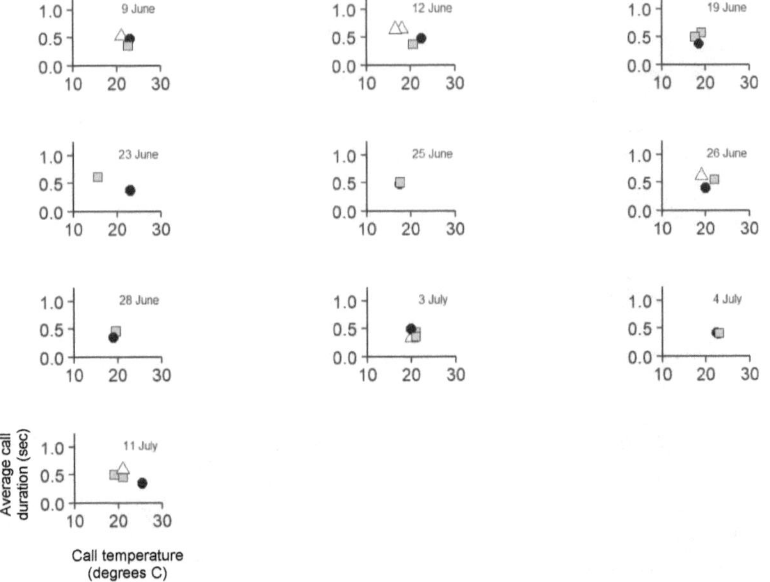

Supplemental Figure 2. Comparison of average call duration and call temperature of male eastern gray treefrogs on nights during the study when more than one male was recorded using a pool in the Tamar Pond field. Males were recorded among those pools with no heterospecifics (filled circles), ten green frog tadpoles (gray squares), or ten bullfrog tadpoles (open triangles). Individual points represent the average call duration for each recorded male using a pool on the date listed.

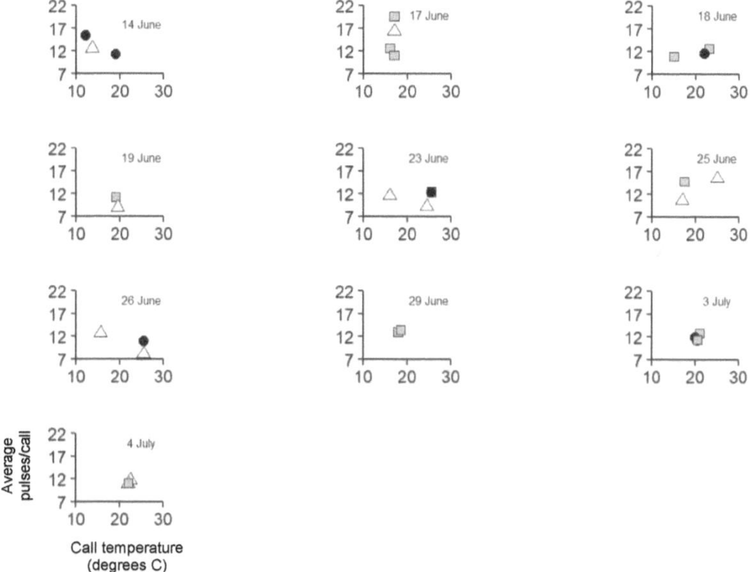

Supplemental Figure 3. Comparison of average number of pulses per call and call temperature of male eastern gray treefrogs on nights during the study when more than one male was recorded using a pool in the Skating Pond field. Males were recorded among those pools with no heterospecifics (filled circles), ten green frog tadpoles (gray squares), or ten bullfrog tadpoles (open triangles). Individual points represent the average number of pulses per call for each recorded male using a pool on the date listed.

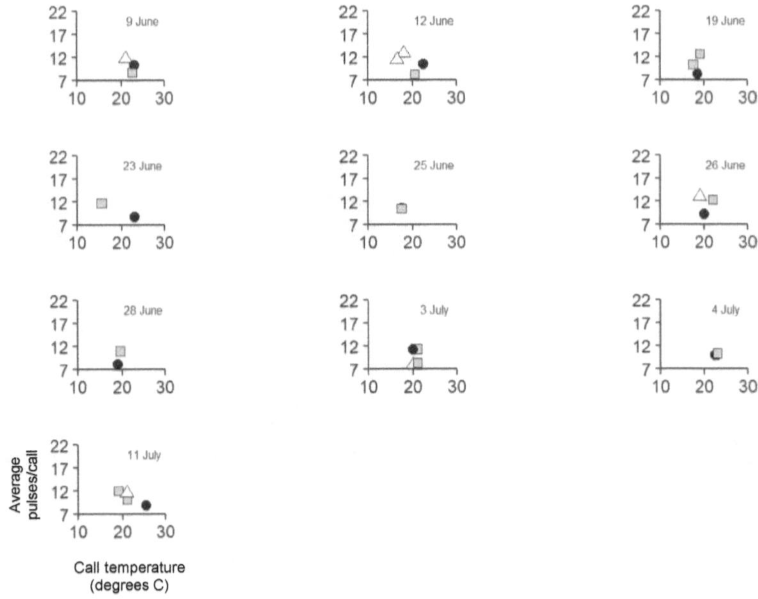

Supplemental Figure 4. Comparison of average number of pulses per call and call temperature of male eastern gray treefrogs on nights during the study when more than one male was recorded using a pool in the Tamar Pond field. Males were recorded among those pools with no heterospecifics (filled circles), ten green frog tadpoles (gray squares), or ten bullfrog tadpoles (open triangles). Individual points represent the average number of pulses per call for each recorded male using a pool on the date listed.

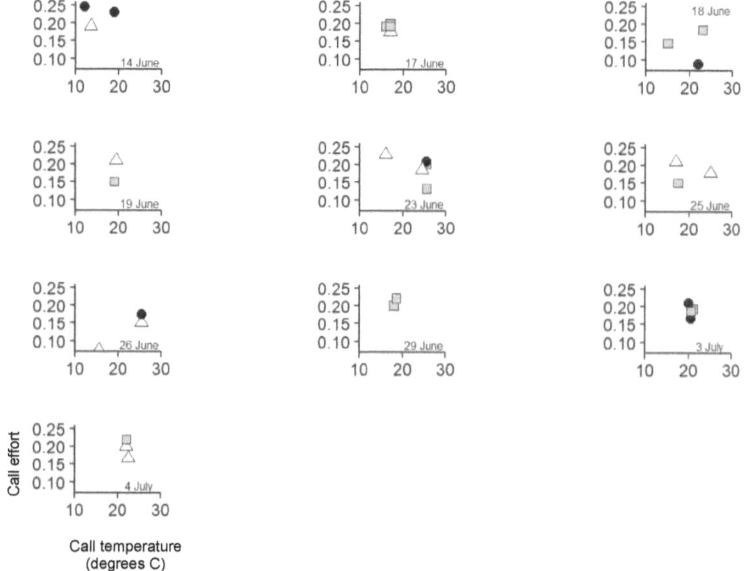

Supplemental Figure 5. Comparison of call effort and call temperature of male eastern

gray treefrogs on nights during the study when more than one male was recorded using a

pool in the Skating Pond field. Males were recorded among those pools with no

heterospecifics (filled circles), ten green frog tadpoles (gray squares), or ten bullfrog

tadpoles (open triangles). Individual points represent the call effort for each recorded

male using a pool on the date listed.

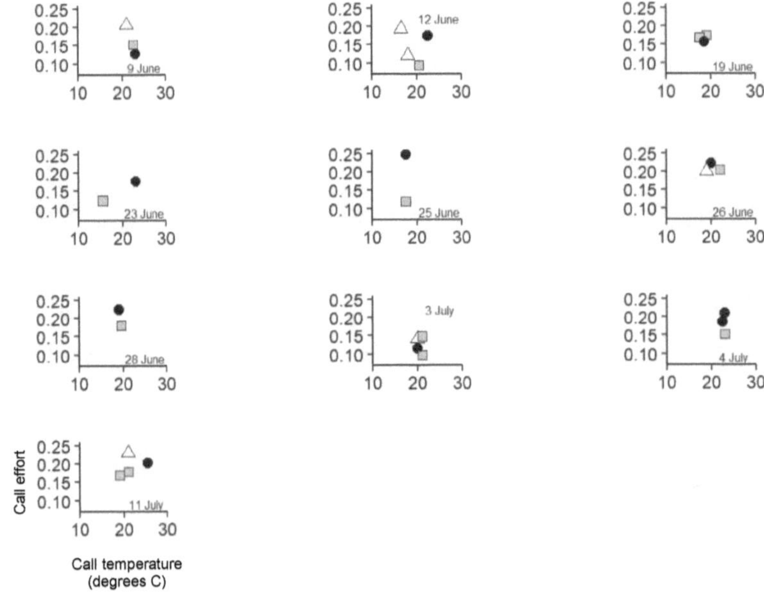

Supplemental Figure 6. Comparison of call effort and call temperature of male eastern

gray treefrogs on nights during the study when more than one male was recorded using a

pool in the Tamar Pond field. Males were recorded among those pools with no

heterospecifics (filled circles), ten green frog tadpoles (gray squares), or ten bullfrog

tadpoles (open triangles). Individual points represent the call effort for each recorded

male using a pool on the date listed.

Chapter 3: Fringe habitat use by male treefrogs: making the best of a bad situation or an adaptive mating strategy?

Authors: David A. Dimitrie[1], Michael F. Benard[1]

1. Department of Biology, Case Western Reserve University, Cleveland, OH

INTRODUCTION

Animals in heterogeneous landscapes must make decisions about which habitats to use, and determining which habitats to use over others for activities such as foraging and breeding can be influenced by both the quality of the available habitats and the resulting fitness outcomes (Schmidt et al. 2010, Abdulwahab et al. 2019, Jourdan et al. 2021). When presented with multiple habitats to choose from, costs and constraints must be accounted for by an animal when determining if it should use a certain habitat over another (Acker et al. 2017). However, habitat selection is not exclusively an adaptive process, and instead mismatches may occur between selected habitats and fitness outcomes (Chalfoun and Schmidt 2012). This can be especially true for selection among nesting and breeding habitats (Gilroy et al. 2011).

The presence of fringe habitats (i.e., those on the periphery of or adjacent to core habitats) on a landscape provide additional habitat choices for animals actively seeking out a breeding location. The state of an animal (e.g., pre- versus post-breeding; disperser versus non-disperser) can influence habitat selection, especially when moving among multiple available habitats (Matthysen 2012). For example, the use of fringe habitats may result from the direct dispersal of individuals from the core habitat (Bowler and Benton 2005), or these habitats may be encountered as a part of prospecting behavior or foraging away from the core habitat (Ponchon et al. 2013). Prospecting of, and ultimately the settlement at, fringe habitats is an important dispersal strategy that is documented in a variety of taxa, most notably birds (Reed et al. 1999). Importantly, these fringe habitats may be characterized by different conditions than an established, core habitat, including

differences in predation, competition, and resources. In some instances, these conditions

may be more favorable than the conditions of the core habitat (Schmidt et al. 2010).

Therefore, certain individuals may be more likely to select these fringe habitats over the

core habitats based on these conditions. It is also possible that novel or fringe habitats

may serve as ecological traps (Schlaepfer et al. 2002, Robertson and Hutto 2006). Thus,

understanding the characteristics of individuals choosing to use these fringe habitats is

important for understanding broader ecological patterns in the context of habitat selection

(Morris 2003).

The probability that an individual becomes a disperser to fringe habitats can be

influenced by a variety of factors across life-history stages (Reed et al. 1999, Benard and

McCauley 2008). Body size and body condition are two proximate factors that can

influence dispersal. The quality of the habitat experienced during early development can

affect dispersal ability and propensity, resulting in larger individuals dispersing more than

smaller individuals (e.g., Baines and McCauley 2018). In anurans, a positive relationship

between dispersal abilities to surrounding aquatic habitats and body size exists during the

breeding season for adult green frogs (*Rana clamitans;* Searcy et al. 2018). In contrast,

male eastern gray treefrogs (*Hyla versicolor*) tend to remain near an established breeding

pond during the breeding season, and those males moving to habitats an average of 30 m

away are no larger nor of higher body condition indices than males remaining closer to or

at the core breeding pond (Johnson et al. 2007). Individuals using fringe habitats may

therefore be phenotypically superior to the mean traits of the core population, or they

may be no different or possibly even inferior to this core population. Understanding if individuals moving to and utilizing fringe breeding habitats are different from core habitat individuals can therefore have important implications for both individual fitness outcomes and resulting population and community dynamics in these fringe habitats.

Anurans are an excellent study system to address these questions of habitat use between core and fringe habitats. Specifically, we used the eastern gray treefrog (*Hyla versicolor*) as our focal species because males use advertisement calls to attract females and they have a prolonged breeding season which allows males to use multiple potential habitats during the season. By producing advertisement calls to attract females, a male should produce the best call necessary at his breeding habitat. Previous work has demonstrated that males will use fringe breeding habitats to advertise from (Chapter 2, this book) and that these males are successful in attracting females to these habitats (Chapters 1 and 2, this book). However, during this previous work we anecdotally documented that males using the fringe habitats typically produced shorter calls than males using the core habitat. Males producing shorter calls are less successful in attracting a female, because a shorter call is an indication to a female treefrog that the male is an inferior sire compared to a male producing a longer call (Gerhardt et al. 1996, Welch et al. 1998). We therefore wanted to test if the calls of males using these fringe breeding habitats differed in attractiveness to females compared to males that use an adjacent core breeding habitat. Although males will use these fringe habitats, these males may be inferior to core habitat males in terms of body size and call attractiveness. Mating with inferior males at these fringe locations may put females at a disadvantage compared to those breeding with residents at the core pond, while breeding with potentially superior males in fringe habitats that are suitable for offspring would be advantageous for females.

Additionally, if there are differences in the calls produced by males using fringe and core habitats, we wanted to test if these differences are static (i.e., due to fixed differences in males) or plastic (i.e., due to the social environment). Males calling in the presence of neighboring calling males increase their call duration compared to when calling in isolation (Wells and Taigen 1986). Therefore, males moved from fringe habitats to core habitats may not change call characteristics because they are inferior to resident males at the core breeding site. Alternatively, males from fringe habitats may change their call characteristics at the core breeding pond to match the social environment and become more attractive in the presence of potential male competitors (and thus they are not inferior to males using the core habitat).

To test if males using fringe breeding habitats are different from males using an established, core breeding habitat, we established an array of artificial aquatic habitats adjacent to a known gray treefrog breeding pond and evaluated how body condition and advertisement calls of the males using these fringe habitats compared to those males using the core habitat. This allowed us to test if males using fringe habitats are different than males using core breeding habitats. We then moved individuals from both habitat types to a pool immediately adjacent to the core breeding habitat pond and measured how their call characters changed when in a similar social environment. This allowed us to test if differences in call characteristics between the two habitat types are due to differences in attractiveness between these two male types, or if they are due to males altering their calls to match their social environment.

METHODS

We conducted our study at a known gray treefrog breeding pond (hereafter, "Skating Pond") at Case Western Reserve University's Squire Valleevue and Valley Ridge Farms in Cuyahoga County, Ohio, USA. We setup 260 L aquatic mesocosms (hereafter "pools") as available habitat for breeding treefrogs in a field adjacent to Skating Pond. We placed 10 pools in a row in the field parallel to Skating Pond so that each pool was 5 m from its closest neighboring pool. Distance from fringe habitat pools to the edge of Skating Pond ranged from 30-60 m. We filled all pools on 11 June and 12 June 2020 with ca. 60 liters of water collected from Skating Pond. Water was filtered through a course mesh to avoid introducing any frog eggs or tadpoles from Skating Pond into the fringe habitat pools. We also setup an additional pool on 11 June located 3 meters from the Skating Pond shore to serve as a core habitat testing pool. We left the pools in the fringe habitat and core habitat testing area until 24 July, at which time we concluded the experiment.

For our research approach, we categorized three sites: Skating Pond represented the "core habitat"; the 10 pools in the field adjacent to Skating Pond represented the "fringe habitat"; and the additional pool 3 m from Skating Pond served as a neutral recording test pool ("core habitat testing pool") immediately adjacent to the core pond that provided a standardized test area at Skating Pond (Figure 1). We checked the fringe habitat for calling males on 22 nights throughout the breeding season. We located 3 unique male frogs using the fringe habitat pools on multiple nights for advertising throughout the study and categorized these males as "fringe residents". We also searched Skating Pond for calling males on most survey nights to locate males using the core habitat for breeding. We located 10 unique male frogs using the core habitat and used 7 in our recording trials. We categorized these males as "core residents". For each individual frog, we first recorded its advertisement call where it was originally found (core habitat or fringe habitat). We then immediately moved that individual to the core habitat testing pool and recorded the animal again. This allowed us to test how each male changed his advertisement call characteristics when moved from his original capture location to the core habitat testing pool. We moved core residents to the core habitat testing pool as a control to test if moving males from their calling location in the main chorus at Skating Pond to the core habitat testing pool changed their call characters. All animals were then released back at their original capture location at the end of each night.

We recorded advertisement calls from each male using a Marantz PMDF 661 field recorder with a Sennheiser ME66/K6P shotgun microphone. We set the microphone on a tripod boom microphone stand approximately 0.5 m from the focal male, began recording, and left the recording area to eliminate any influence on the male's advertisement call. We then allowed the recording to proceed for at least 3 minutes when possible. We took dorsum photos of each frog and measured body mass and snout-vent length (SVL) following recording to eliminate the effects of animal handling on call characteristics. We used a handheld infrared thermometer (Extech Instruments) to measure the temperature of the frog immediately following the recording. For the 3 additional core residents we located but did not record, we also measured body mass and SVL. For all recordings, we calculated average call duration, average number of calls produced, and overall call effort (proportion of time spent calling) on a 1-minute portion of each recording (generally during the middle of each recording when frog calling behavior was stable). Call characters were measured using Raven Pro 1.6 software (Center for Conservation Bioacoustics, Cornell Lab of Ornithology).

Statistical analyses

To investigate differences in frog body condition between those males using fringe habitats and the core habitat, we used an ANCOVA with SVL as a covariate and resident type as a fixed effect on frog mass. We assigned each individual male used throughout the study a resident type (core or fringe) based on which habitat type he was first found in during the breeding season (n=3 fringe habitat males, n=10 core habitat males). For fringe habitat males, we calculated the mean mass and SVL measured from all captures for that individual frog since each of these 3 males were used on multiple nights. We then tested if frogs calling from fringe habitats produced calls that were different than frogs calling at the core habitat using a series of linear mixed-effects models. We tested whether the original habitat location (fringe or core) affected average call duration, number of calls produced per minute, and call effort. For each call character, we included the call character of the frog at his original capture location in either the fringe or core habitats as the response variable, and original capture location (fringe or core), calling temperature, and their interaction as fixed effects in each model. We also included frog ID as a random effect (1|frog ID) in the models because the 3 males using the fringe habitats throughout the season were used on multiple nights for call character analyses. We included calling temperature in our models because it can affect advertisement call characters in our study population (Chapter 2, this book). Models were created with the *lmer* function in the *lme4* package in R.

We used paired t-tests to test if the 3 fringe habitat males changed their call characters when moved from the fringe habitat to the core habitat. We calculated the change in call characters for each male on each night when he was moved from the fringe habitat to the core habitat (change in character = call character in core habitat testing pool - call character in fringe habitat). We then calculated the mean change in these values for each of the three males for all nights he was recorded (3, 6, and 2 nights for male #1, 2, and 3, respectively) and used these mean values in a paired t-test for each of the call characters. We also conducted paired t-tests using the change in call characters for core habitat males when they were moved to the testing pool adjacent to the core habitat (call character in core habitat testing pool - call character at original capture location in pond). Each core male was only recorded on one night, so we used the difference in values for each call character on the single night that male was tested. We used these paired t-tests for core habitat males as a control to assess how moving males from their calling location in the main chorus to the core habitat testing pool changed their call characters. We also conducted a t-test on each of the three call characters between fringe and core habitat males when in the core habitat testing pool to test if their calls differed in this common environment. We conducted all analyses using R software environment version 4.0.5 (R Core Team 2021).

RESULTS

We located 3 unique male frogs using the fringe habitat pools for advertising throughout the study. These males were located and recorded at both their original fringe habitat and the core habitat testing pool on 2, 4, and 7 nights each. On any given night, these males were never recorded in the fringe habitat when calling closer than 10 m from the closest calling male. We located 10 unique male frogs using the core habitat. We recorded 7 of these males on one night each at both their original advertising location in Skating Pond and in the core habitat testing pool. There was no interaction between original capture location (fringe or core) and SVL on frog mass ($F_{1,9} = 0.21$, $P = 0.66$) nor was there was a difference in frog mass between the two original capture locations ($F_{1,9} = 1.35$, $P = 0.28$), indicating that males using the two habitat types did not differ in overall body condition (Figure 2).

Call traits at original locations

Resident type significantly affected both average call duration and number of calls produced for males at their original capture location (Table 1). The fringe habitat males produced calls that were an average of 0.24 seconds shorter than males calling from the core habitat but produced 9 more calls per minute on average than males calling from the core habitat (Figure 3A, 3B). Call effort was similar between both resident types (Figure 3C). Neither call temperature nor the interaction between resident type and call temperature affected any of the call characters (Table 1).

Change in call traits when moved to core habitat testing pool

The fringe habitat males increased their call duration ($t_2 = 27.7$, $P = 0.001$) and decreased their number of calls ($t_2 = 12.9$, $P = 0.006$) when moved from the fringe habitat to the core habitat, resulting in an overall similar call effort in both habitat types ($t_2 = 0.85$, $P = 0.49$; Figure 4). The core habitat males, however, did not significantly change their average call duration ($t_6 = 1.98$, $P = 0.09$), number of calls ($t_6 = 1.37$, $P = 0.22$), or call effort ($t_6 = 2.32$, $P = 0.06$) when moved from their original location in the core habitat to the core habitat testing pool (Figure 4). The changes in fringe male calling traits when moved to the testing pool resulted in there being no difference in call traits between fringe males and core males in the testing pool. Average call duration ($t_8 = 0.22$, $P = 0.83$), number of calls ($t_8 = 0.84$, $P = 0.44$), and call effort ($t_8 = 0.85$, $P = 0.42$) in the core habitat testing pool were the same between both resident type males (Figure 5).

DISCUSSION

We set out to assess if organisms using fringe breeding habitats differ from conspecifics using an established, core habitat because decisions on which habitat to utilize during the breeding season can affect an organism's fitness (Schmidt et al 2010). Specifically, we tested if males of the eastern gray treefrog that use fringe breeding habitats are different than males using an established breeding pond. We evaluated body condition and advertisement calls of males using the fringe habitats and compared them to those of males using the core habitat. Although we only documented three frogs using our fringe habitat throughout the study, frogs using the fringe habitat were of similar body condition to those males using the core habitat. However, although they called at a similar overall effort as the core habitat males, fringe habitat males produced more calls per minute and shorter calls than males in the core habitat. When fringe animals were moved to the core habitat, they adjusted their calls to match those produced by the males at the core habitat.

A central question to our study was why do males use these fringe habitats over the core breeding pond? Males that selected fringe habitats may have encountered these pools during their breeding migration to the core habitat pond and remained to advertise to females. Alternatively, these males may have moved away from the core breeding pond to prospect additional habitats or forage and may have encountered the fringe habitats. Although we cannot specifically determine if the fringe males were dispersers, prospectors, or simply individuals that located the fringe habitat pools during their breeding migration to the core habitat, our results demonstrate that the males using the fringe habitat were similar in both body condition and advertisement call ability to the males we tested using the core habitat.

Importantly, we documented that fringe males adjust their advertisement calls accordingly to account for the lack of nearby competing males while in the fringe habitat and to account for the presence of nearby males within the natural pond chorus when at the core habitat. These results support previous work by Wells and Taigen (1986) demonstrating that males calling within a chorus of other males increase their call duration and decrease the number of calls they produce compared to when calling in isolation. This results in an overall similar call effort and a resulting constant level of aerobic metabolism and oxygen consumption between males in the two habitat types. Wells and Taigen (1986) hypothesized that that production of long calls leads to a more rapid depletion of glycogen in the muscles that are used during calling. Therefore, the production of short calls by males in the absence of other calling males reduces muscle glycogen depletion. This would increase overall calling endurance for fringe males. Thus, although males in the fringe habitat produce calls which are less attractive to females, the overall increase in calling endurance would allow a male to advertise on more nights throughout the breeding season. In many observational and experimental field studies, the number of nights spent advertising is a better predictor of male mating success than advertisement call characters or body condition in many species of frogs, including *H.*

versicolor (Gerhardt et al. 1987; Sullivan and Henshaw 1992, Bertram et al. 1996). However, once in the presence of competing males such as at the core habitat, competition for females is high and thus a more attractive call is important even if the overall time a male can spend calling is reduced. This plasticity in signal features has been well demonstrated in the context of treefrog advertisement calls in a variety of social interactions when at a breeding location (e.g., Neelon and Höbel 2019).

Our results support the hypothesis that signal plasticity may be an adaptive strategy when choosing among multiple, isolated habitats. Signal plasticity allowed males to still be successful when calling from the fringe habitats, even if their calls would be perceived as of lesser quality to females. Although we did not observe any of our three fringe habitat males in amplexus with females on any of our survey nights, males were successful at obtaining a mate in these fringe habitats. We documented oviposition events (eggs found the following day) in four different pools during this study. Oviposition and direct evidence of males obtaining a female (males in amplexus with a female) have also been observed previously in our fringe habitat pools (Chapters 1 and 2, this book). Therefore, it may be beneficial for males to utilize these fringe habitats without competing males present in order to increase their chance of obtaining a mate. Treefrogs in the fringe habitats may also be adopting the win-stay/lose-switch habitat selection strategy (Campomizzi et al. 2012), where successfully establishing a calling location and attracting a female results in a male staying at a fringe habitat, while an unsuccessful male may move among multiple fringe habitats or move to the core habitat in an attempt to increase his mating success. Each of the three fringe males were found at multiple locations within the fringe habitat throughout the breeding season. These males may have used different habitats over multiple nights to potentially increase their mating success. This study, along with Chapter 2, also raises the possibility that core males may move to fringe habitats during the breeding season as part of the win-stay/lose-switch strategy. Although we did not document any change in the identity of the males using the fringe habitats during this study, it is possible that core males may move to fringe habitats during the season if they are unsuccessful at the core habitat. In Chapter 2, we documented new males colonizing our fringe habitat pools throughout the breeding season. However, we cannot be certain if these were males that moved from the core habitat or were males that were initially migrating to the core habitat and instead selected the fringe habitats they encountered while moving to the core habitat. Similarly, some males using our fringe habitat pools in Chapter 2 early in the breeding season may have moved to the core habitat because many of the males we found early in the breeding season in Chapter 2 using our pools were no longer present by the end of the study. However, not all males will actively breed throughout the entire season, and some may arrive first to a breeding site while others later (Fellers 1979). These hypotheses as to why males may begin using fringe habitats at different times throughout the breeding season would be an important follow-up to both this study and that of Chapter 2.

Body size and body condition (mass for a given size) are traits that can represent fitness (Peig and Green 2009). Therefore, examining the relationship between an individual's size and his use of fringe versus core habitat can help determine if males using fringe habitats are of different quality than those using the core habitat. We found no difference in the body mass or body condition between the two resident type males. However, other size metrics can be linked to dispersal abilities in anurans. For example, cane toads (*Rhinella marina*) with longer legs move faster and disperse farther than individuals with shorter legs (Phillips et al. 2006), and dispersing male wood frogs *(Rana sylvatica)* are larger than non-dispersers (Berven and Grudzien 1990). Searcy et al. (2018) also found that dispersing green frogs (*Rana clamitans*) are larger than non-dispersers, but limb-length did not affect dispersal. Interestingly, they also found that dispersers, although larger in snout-vent length, tended to have lower body condition indices. The authors hypothesize that this may be because dispersal is energetically costly and thus results in a reduction in body condition that is then present at the new habitat. Alternatively, it may be that individuals with lower body conditions are poor competitors in the original habitat and are thus more likely to disperse to a new habitat as a competitive release mechanism. In general, larger individuals are more likely to be dispersers to new habitats than smaller individuals across taxa including insects, amphibians, and birds (Benard and McCauley 2008, Camacho et al. 2019, May 2019), but this may not be the case for individuals using fringe habitats that are only a certain distance away from a core habitat. Johnson et al. (2007) documented that most male eastern gray treefrogs in a population tend to remain near a core breeding pond during the breeding season, and those males moving to surrounding habitats an average of 30 m away are no larger nor of higher body condition indices than males remaining closer to or at the core breeding pond. It is also important to note that core males in our study were opportunistically captured and used in this study, so size data may be biased towards animals of certain body conditions (for example, if only calling frogs of smaller size at Skating Pond were able to be captured while larger frogs called from inaccessible locations). Future work at Skating Pond will focus on collecting a larger number and more representative sample of the calling male population to compare to males using the fringe habitat pools and other surrounding fringe habitats.

Individuals may be able to use different habitat selection strategies to optimize fitness (Chalfoun and Martin 2007). Our evidence suggests that males using fringe habitat are neither inferior nor superior to those using the core habitat based on their body size, body condition, and call plasticity in the core habitat testing pool. Male treefrogs may use fringe habitats as a method to be able to increase their breeding longevity (i.e., spend more total nights calling) in the absence of competitors at the cost of reducing their per-night chance of attracting a female. This may result in a similar fitness as if the males spent fewer nights calling at the core habitat amongst competing males, or perhaps the use of fringe habitats for these three males is a strategy that does optimize their individual fitness. Additional tests are needed to assess if these males ultimately increase their mating success in fringe habitats compared to the core habitat.

TABLES

Table 1. Linear mixed-effects model results for original capture location (core or fringe) and calling temperature effects on the original call characters of call duration, number of calls produced per minute, and call effort recorded from male gray treefrogs using either pools in the fringe habitat or using the core natural pond habitat.

Call character	Factor	χ^2	P
Call duration	Original capture location	**5.13**	**0.023**
	Call temperature	3.64	0.06
	Original capture location * Call temperature	0.09	0.76
Number of calls	Original capture location	**13.07**	**<0.001**
	Call temperature	0.02	0.89
	Original capture location * Call temperature	1.37	0.24
Call effort	Original capture location	0.01	0.92
	Call temperature	1.45	0.23
	Original capture location * Call temperature	2.64	0.10

Figure 1. Map of the study area and the three sites utilized at Case Western Reserve University's Squire Valleevue and Valley Ridge Farms: Core Habitat (Skating Pond), Fringe Habitat, and the Core Habitat Testing Pool. Blue circles in the Fringe Habitat represent the 10 artificial pools placed to serve as breeding habitats for fringe males. Blue circle at the Core Habitat Testing Pool represents the artificial pool used to test call characters of males in a standardized test area at the core habitat.

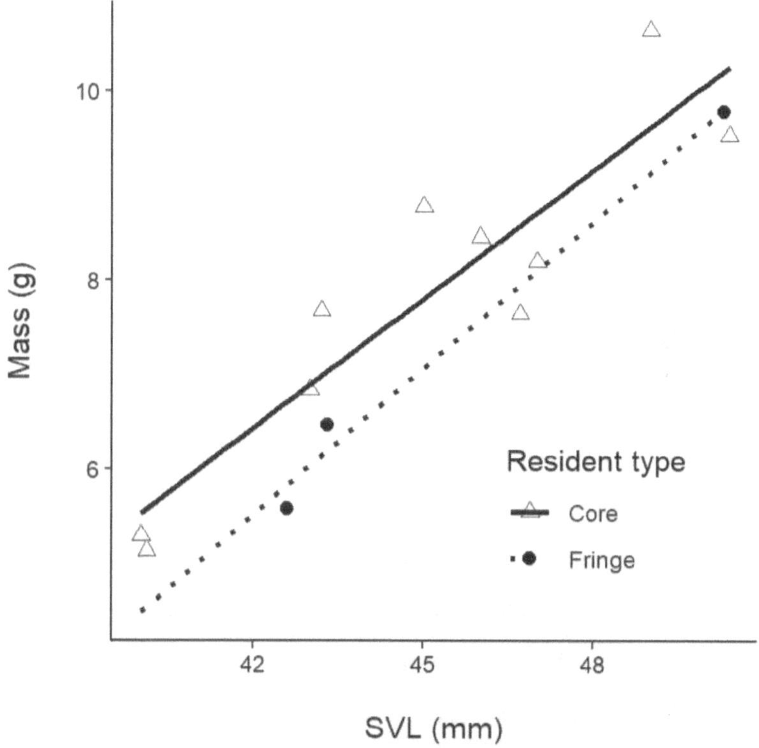

Figure 2. Body condition (relationship between snout-vent length (SVL) and mass) of frogs using the core and fringe habitat types. Each individual point for a resident type represents the SVL and mass measurement for an individual frog used during the study. For the three frogs located and recorded on multiple nights using the fringe habitat type, values are mean SVL and mass for all captures over the entire study.

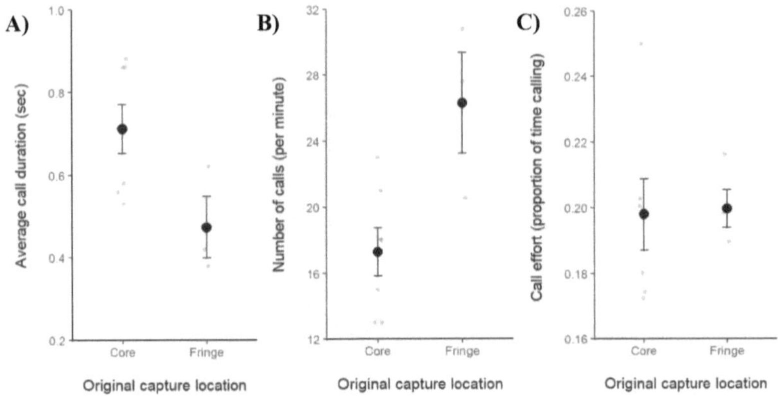

Figure 3. Call characters of advertising male treefrogs at their original capture location in the two habitat types (core or fringe). A) Average call duration, B) number of calls per minutes, and C) call effort. Data are mean ± S.E.M. for each resident type. Individual points represent data for each individual male at the original capture location (for fringe males, values are the average over all recordings for that male to account for the use of recordings on multiple nights from each male within this resident type).

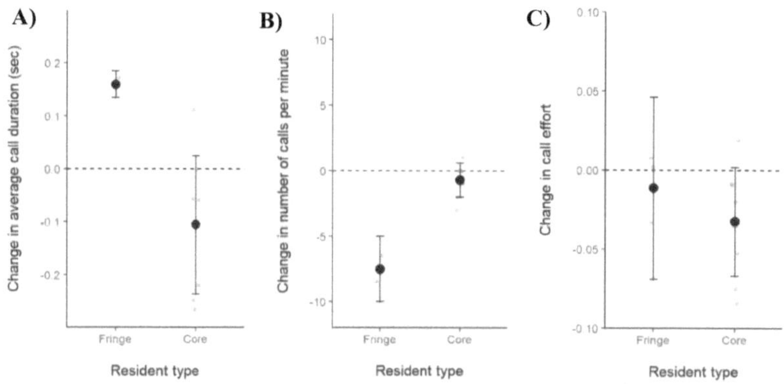

Figure 4. Change in call characters between males of both resident types when moved from their original capture location (i.e., in the fringe habitat pools or at Skating Pond) to the core habitat testing pool adjacent to Skating Pond. A) Change in average call duration, B) change in number of calls produced per minute, and C) change in overall call effort (proportion of time spent calling). Changes in each call character were calculated as change in character = call character in core habitat testing pool - call character at original capture location. Dashed line represents no change for each respective call character. Data are mean ± 95% C.I. for each resident type. Individual points represent data for each individual male within each resident type (for fringe males, values are the average change in each call character over all recordings for that male to account for the use of recordings on multiple nights from each male within this resident type).

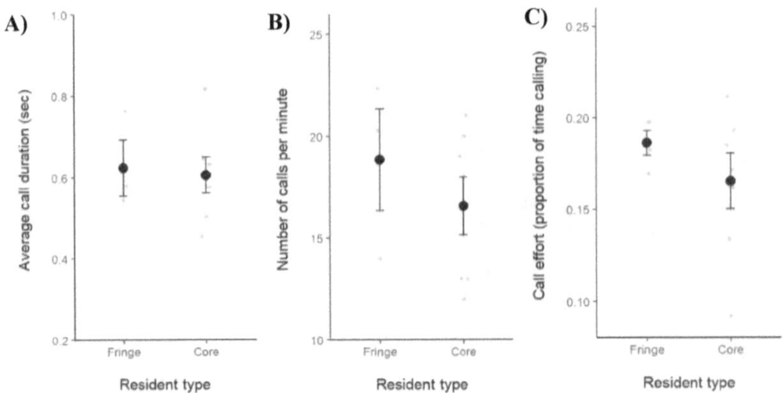

Figure 5. Call characters between males of both resident types when in the core habitat testing pool adjacent to Skating Pond after being moved from their original capture location (i.e., in the fringe habitat pools or at Skating Pond). A) Average call duration, B) number of calls produced per minute, and C) overall call effort (proportion of time spent calling). Data are mean ± S.E.M. for each resident type. Individual points represent data for each individual male within each resident type (for fringe males, values are the average values for each call character over all recordings for that male to account for the use of recordings on multiple nights from each male within this resident type).

Chapter 4: Response of American toads and their invertebrate prey to experimentally elevated soil pH

In revision at Ichthyology and Herpetology

Authors: David A. Dimitrie[1], David J. Burke[2], and Michael F. Benard[1]

[1]Department of Biology, Case Western Reserve University, Cleveland, OH

[2]Holden Arboretum, Kirtland, OH

INTRODUCTION

Acid deposition resulting from anthropogenic activities, predominantly the combustion of fossil fuels since the Industrial Revolution, has led to negative environmental consequences. Forests in temperate regions of Europe and North America that are not characterized by naturally acidic soils or a strong neutralizing capacity can be especially sensitive to acidification (Long et al., 2009; Burns et al., 2011). Acidification can have broad effects on forest floral composition and health, negatively affecting a diverse array or organisms including soil biota (Kuperman et al., 2002), mammals (Pabian et al., 2012), and birds (Hames et al., 2002). Although acidified habitats may directly affect organisms, effects may also occur indirectly due to changes in interspecific interactions and trophic dynamics (Graveland et al., 1994; Mänd et al., 2000). Amphibians are especially sensitive to acidification, and many studies have documented negative effects of acidification on the embryonic and larval stages due to the close association of these life stages with aquatic habitats (see Pierce, 1993) Survival and growth of larval frogs and salamanders are often negatively affected in acidic habitats due to direct physiological effects (Brodman, 1993; Rowe et al., 1992). Acidification can also alter interspecific interactions among amphibians in the aquatic environment through indirect, non-lethal effects on predator size, which can alter predation success and

ultimately community dynamics (Kiesecker, 1996). However, the effects of acidification

on post-metamorphic amphibians and their interspecific interactions in terrestrial

environments have received far less attention than the aquatic embryonic and larval

stages.

Amphibian diversity is often lower in hardwood forests where habitats have been

acidified (Wyman and Jancola, 1992), but how acidification directly and indirectly

affects specific species is less well known. This knowledge gap is important to address

because many amphibians inhabit terrestrial habitats following metamorphosis and

emergence from aquatic habitats, and the implications of terrestrial habitat management

are often overlooked in amphibian conservation (Dodd, 2010). To date, most work

examining the terrestrial response of amphibians to acidification has been conducted on

salamanders. Acidic forest floor soil can disrupt sodium balance in several species of

terrestrial salamanders (Frisbie and Wyman, 1991), which can lead to changes in their

density and distribution in forest ecosystems (Wyman and Hawksley-Lescault, 1987).

Marbled Salamanders (*Ambystoma opacum*) are negatively affected by acidified soils

following metamorphosis, both directly through mortality at a pH of 4.0 and via lower

growth at a pH of 5.0 (Anderson and Johnson, 2018). However, sensitivity to acidic

conditions varies among species and populations in terrestrial environments (Pierce,

1993). Recently, there has also been considerable work demonstrating high levels of

intraspecific variation in salamander tolerance to acidified habitats. For example, certain

populations of the Red-backed Salamander (*Plethodon cinereus*) are tolerant of acidified

soils and display no differences in individual body condition or population density in

acidified forest habitats compared to more pH neutral forest habitats (Moore and Wyman,
2010; Cameron et al., 2016). This intraspecific variation in soil pH tolerance in Red-
backed Salamanders is due to local adaptation capabilities as well as a generalist feeding
strategy, allowing populations to persist in habitats with pH values below 3.8 (Bondi et
al., 2016; Bondi et al., 2019). In addition to salamanders, the presence and distribution of
certain anurans such as the American toad (*Anaxyrus americanus*) are restricted in some
forests to locations with soil pH above 4.0 compared to other more acid-tolerant species
such as the Wood Frog (*Rana sylvatica;* Wyman, 1988). However, direct experimental
evidence comparing the response of anurans in acidified terrestrial habitats to locations
less affected by acidification is lacking. Overall, it is still unclear which species of
amphibians are more sensitive than others to acidified terrestrial habitats and if this
sensitivity is driven by either direct (e.g., physiological) or indirect (e.g., prey-mediated)
effects.

In addition to lethal and non-lethal effects on terrestrial amphibians, acidified
habitats have the potential to alter trophic dynamics. Amphibians play an important role
as consumers in shaping the composition of forest floor invertebrate communities, so
changes in amphibian populations linked to acidification would alter trophic interactions
and ultimately affect ecosystem processes (Davic and Welch, 2004). For example, the
Red-backed Salamander can reduce the decomposition rate of leaf litter through top-
down consumer effects on leaf litter fragmenters and thus alter forest floor community
function and carbon dynamics (Wyman, 1998). Therefore, effects of long-term
acidification on consumers such as amphibians could translate to top-down effects on
these invertebrate communities. Conversely, effects of acidification on invertebrate

communities may have important bottom-up trophic effects on consumers such as forest floor amphibians. Invertebrate abundance, for example, may be lower in acidified forests (Kuperman, 1996), and decreases in key invertebrate prey for amphibians may therefore lead to indirect negative effects of acidification on amphibians. Understanding the effects of acidification on the components of forest faunal communities, including both amphibians and invertebrates, is thus important for the conservation of these groups of organisms and can assist managers with decisions regarding conservation objectives in acidified forests.

We examined the effects of acidification on a common North American amphibian and the leaf litter invertebrate community on which it subsists using experimental forest plots that either have had soil pH raised to pre-acidification levels via addition of Hi-Ca lime or that still have acidified soils. This contrast allowed us to test how the forest floor community responds to soil acidification by producing conditions that mimic pre-acidification. We housed American toads in terrestrial enclosures in either untreated forest plots with acidified soils or forest plots with experimentally elevated soil pH and evaluated toad survival, growth, and diet. We also conducted pitfall sampling to assess invertebrate abundances. Our aim was to answer two questions. First, does soil acidification affect post-metamorphic American toad survival and growth? If acidic soils negatively affect American toad survival and growth due to either direct physiological effects or indirectly due to a negative effect on prey (i.e., invertebrate) availability, then American toad survival and growth would be higher in forest plots treated with experimentally elevated soil pH. Second, does soil acidification affect the forest floor invertebrate community? If acidification negatively affects the invertebrate community, then invertebrate abundances and overall invertebrate diversity would be higher in forest

plots with elevated soil. We also tested whether American toads regulate the forest floor

invertebrate community. If American toads exert a top-down effect on the invertebrate

community, then invertebrate abundances would be lower in enclosures with toads

present compared to enclosures without toads present due to top-down predation by

toads. Invertebrate abundances and toad diet analyses at the end of our study also allowed

us to examine if American toads acted as diet generalists or preferentially selected certain

prey and thus influenced the structure of the forest floor invertebrate community.

METHODS

Study area

We conducted our research within three mixed hardwood forests in northeast

Ohio. Two of these forests are located at the Holden Arboretum, Lake County, OH

(Pierson Creek Forest: 41°36'38"N, 81°18'39"W, elevation 303 m; Schoop Forest:

41°36'44"N, 81°19'15"W, elevation 310 m) while the third is located at Case Western

Reserve University's Squire Valleevue and Valley Ridge Farm (hereafter "CWRU

Farm") in Hunting Valley, Cuyahoga County, OH (41°29'60"N, 81°25'15"W, elevation

325 m). The distances between forests are at least 1 km but no greater than 15 km. These

three forests are characteristic of hardwood forests throughout eastern North America,

composed primarily of *Fagus grandifolia* (American Beech), *Quercus* spp. (oak), and

Acer spp. (maple). These forests were selected because they have been studied since 2009

as part of a forest-level pH experiment (Kluber et al., 2012). Each forest consists of three

800 m^2 (20 x 40 m) plots that have been managed via the application of Hi-Ca lime to

elevate soil pH. An additional three untreated plots with historically acidified soils are located within each forest. Hi-Ca lime was applied to all treatment plots initially in the fall of 2009, 2010, and 2012. Subsequent treatment has occurred since 2012 on an as-needed basis. This has resulted in an elevated soil pH range of 5.8 – 6.2 for lime-treated plots compared to 4.1 – 4.6 for untreated plots.

Study design

We crossed two soil pH treatments (untreated, acidic soil pH and lime-treated, elevated soil pH) with three enclosure types (enclosures with American toads, enclosures without American toads, and no enclosures) in a 2x3 design. The two soil pH treatments allowed us to compare toad and invertebrate community responses between forest plots with elevated soil pH and plots with acidified soils. We established three 1 m^2 subplot types consisting of different enclosure setups within each forest plot to address several objectives. The enclosures with toads were used to compare toad growth and survival in forest plots with elevated soil pH to forest plots with acidic soil pH. The enclosures without toads allowed us to evaluate how the presence or absence of toads affected the invertebrate community by comparing invertebrate abundances from subplots with and without toads. The subplot without an enclosure was an additional methodological control to assess how the presence or absence of a constructed enclosure may affect the invertebrate community. Subplots were randomly located along one of the long edges of the study plot perimeter, and subplots were grouped so that each was approximately 2 m from the adjacent subplot (Fig. 1). We used this 2x3 design in the three forests and within the three replicate study plots per soil pH treatment located within each forest. In total, we established 54 subplots (3 forests x 2 soil pH treatments x 3 enclosure treatments x 3 forest plots per soil pH treatment per forest).

We collected eggs from three clutches of American toads from a pond at the CWRU Farm and eggs from three clutches from a pond at the Holden Arboretum on 18 April 2017. We used the American toad as our focal amphibian because it is a habitat and diet generalist that feeds on forest floor invertebrates following metamorphosis. We brought all eggs to the CWRU Farm and placed each clutch into an individual nursery box within a 1,100 L aquatic mesocosm. Once tadpoles reached free swimming stage, we haphazardly selected 50 tadpoles from each nursery box and placed each group of 50 tadpoles into a separate 1,100 L aquatic mesocosm based on their population of origin (three mesocosms for the CWRU Farm and three mesocosms for the Holden Arboretum). We raised tadpoles from each site in separate mesocosms so that toad metamorphs from the CWRU Farm pond would later be used in CWRU Farm subplots and toad metamorphs from the Holden Arboretum pond would be used in Holden Arboretum subplots. Aquatic mesocosms were stocked with 10 g rabbit chow and 1 kg of mixed-hardwood leaves to provide food and refugia for the tadpoles. Mesocosms were covered with tightly fitted lids made of 60% shade cloth to prevent colonization by other organisms. Tadpoles remained in mesocosms until completion of metamorphosis (i.e., emergence of all four limbs and complete resorption of tail).

We placed four toads into one subplot with an enclosure within each forest plot (n = 18 enclosures with toads) once metamorphosis was complete. We built terrestrial enclosures in each forest plot using silt construction fencing that measured approximately 1 m in height. The bottom of the fencing was buried 20 cm into the soil and further reinforced with garden staples to prevent toad and invertebrate escapes. A baffle of screen was attached to the top of the fencing to further prevent escape and to discourage other animals from potentially colonizing the enclosure. Enclosure construction was completed between 26 April and 15 May 2017 within all forest plots. Toads from the Holden Arboretum reached metamorphosis one week prior to toads from the CWRU Farm (after 45 and 52 days, respectively). We placed toads into enclosures at the Holden Arboretum between 8 and 12 June 2017 and toads into enclosures at the CWRU Farm on 15 June 2017. We weighed toads following removal from the aquatic mesocosms on the same day they were placed into terrestrial enclosures.

Toads remained in their enclosures for 90 days. We searched each enclosure for five minutes on days 25 and 60 by gently searching the surface of the leaf litter and under leaves. We counted the number of toads located in each enclosure and weighed all toads before releasing them back into their enclosures. We again searched all enclosures on day 90, counted and weighed toads from each enclosure, and then euthanized each toad in MS-222. After euthanasia, toads were fixed in 10% buffered formalin and then stored in 70% ethanol. We later dissected stomach contents from toads captured on day 90 and identified invertebrate contents to class, order, or family when possible to compare diets between the two soil pH treatments (see Invertebrate sampling below). We again searched each enclosure the following day (day 91) and conducted an exhaustive search of all leaf litter on day 92 to ensure that all remaining toads were located. We located no additional toads during these surveys on days 91 or 92, suggesting all remaining toads were located on day 90.

Invertebrate sampling

We sampled invertebrates within all 54 subplots on day 0 and again on day 91. We sampled on both day 0 and day 91 to assess the invertebrate community available as prey to toads both immediately before and after toad presence. We also wished to quantify any change in invertebrate communities due to the presence of toads for 90 days (by comparing invertebrate communities from enclosures with and without toads on day 91). Sampling all 54 subplots also allowed us to compare the invertebrate community from our subplots without enclosures to subplots with enclosures to evaluate the effects of enclosure presence on the forest floor invertebrate community. We placed four 500 ml pitfall traps into the ground within each 1 m^2 enclosure treatment subplot so that each of the four traps were located within one of the four corners of the subplot. Traps were filled with approximately 60 ml of 95% ethanol and left in place for 24 hours (generally from 12:00-12:00). We removed traps after 24 hours and pooled the contents into one sample per subplot. We later identified and quantified invertebrates according to class, order, or family when possible using a dissecting microscope.

Soil pH sampling

Soil pH within each 800 m^2 forest plot has been determined annually as part of an ongoing long-term study within the three study forests (Kluber et al., 2012; Carrino-Kyker et al., 2016). However, we also measured soil pH within each subplot to assess any differences at the macro- (i.e., 800 m^2) and micro-habitat (i.e., 1 m^2 subplot) scale. We collected three soil core sub-samples (depth ca. 5 cm, diameter ca. 5 cm) from each subplot between 19 and 21 September 2017. These sub-samples were homogenized, and 6 ml of deionized water was added to 3 g of homogenized soil from each subplot to create an aqueous mixture (i.e., a 2:1 water:soil mixture). We measured the pH of the resulting aqueous solution using an ExTech ExStik II EC500 pH meter. All pH measurements were taken within 24 hours of soil collection.

Statistical analyses

We tested for differences in soil pH between soil pH treatments using a linear mixed model via the *lmer* function in the *lme4* package in R. The model included soil pH as the response variable, soil pH treatment as a fixed effect, and a random intercept term of subplot nested within forest (1|forest/subplot). We then tested for effects of soil pH treatment, forest, and their interaction on toad survival to day 90 using a generalized linear model. We used the *cbind* function to evaluate our response variable, survival, using the number of individual toads located per enclosure and the number of individuals not located. We used the *glm* function with a quasibinomial error structure to account for overdispersion in the data. We used a repeated-measures ANOVA to test for effects of soil pH treatment and forest on toad mass at each census period (i.e., day 25, 60, and 90). Enclosures represented the experimental unit instead of individual toads; therefore, enclosure means for mass were used in this model. The repeated-measures model included soil pH treatment and forest as between-subject test variables, and time, the interaction between time and soil pH treatment, the interaction between time and forest, and the interaction of all three as within-subject test variables.

We used a series of linear mixed effect models to test for effects of soil pH treatment and enclosure treatment on invertebrate abundances and overall taxonomic diversity (total number of invertebrate groups) on day 0 and day 91. For each invertebrate group, we included the abundance of that group at the sampling period (day 0 or day 91) as the response variable, with the main and interactive effects of soil pH treatment and enclosure type as fixed effects. The models included a random intercept of forest and subplot nested within forest (1|forest/subplot). Models were created using the *lmer* function in the *lme4* package in R. We log transformed or log+1 transformed

invertebrate group data as needed prior to analyses to correct violations in the

assumptions of normality. We also visually inspected diagnostic plots for each model to

assess model fit before interpreting results. These models addressed several aspects of

our research objectives. Inclusion of soil pH treatment in the models tested for

differences in invertebrate abundances between forest plots with elevated soil pH and

acidified soil. Inclusion of enclosure treatment allowed us to test for two other effects on

invertebrate abundances. Any significant effect of enclosure type in our models was

followed by post-hoc pairwise comparisons among the three enclosure types using the

emmeans function in R. First, this allowed us to test for an effect of enclosure presence

by looking at the differences between subplots without enclosures and to subplots

containing enclosures with and without toads. Second, pairwise comparisons among

subplot types also allowed us to test whether invertebrate abundances differed between

enclosures with and without toads to assess if toads affected the invertebrate community

after 90 days.

To further examine differences in overall invertebrate community composition

between soil pH treatments, we used non-metric multidimensional scaling (NMDS) via

the *metaMDS* function in the *vegan* package in R to visualize community patterns

between the two treatments. Community analyses were based on Bray-Curtis distances

and independently completed for both day 0 and day 91 invertebrate data. We pooled

data from all three subplots in each forest plot to create an invertebrate group abundance

matrix for the 18 forest plots. To reduce effects from rare species (those comprising <1%

of either invertebrate samples or toad diet samples), we only included Collembola,

Diptera, Acari, Araneae, Coleoptera, and Hymenoptera. We used the *ordiellipse* function and generated standard error intervals around points based on soil pH treatment type. We then used an analysis of similarities (ANOSIM; *anosim* function, *vegan* library in R) with 9999 simulations to test for a difference in community composition between soil pH treatment types, both on day 0 and day 91.

We used multiple approaches to investigate potential effects of soil pH treatment on toad stomach contents. First, we calculated Pianka's measure of niche overlap (Pianka 1973) between the two treatments based on counts of different types of invertebrate prey. This metric returns values that run from 0 (no overlap) to 1 (complete overlap). Second, we calculated a modified index of relative importance (I_x) based on numerical abundance and relative frequency (Anthony 2008, Bondi et al. 2019), where n_x was the number each type of prey, N was the total number of prey in toad stomachs from that prey group, f_x was the frequency of toad stomachs that contained that prey, and F was the sum of the frequencies.

$$I_x = [(n_x / N) + (f_x / F)] / 2$$

For the four most common invertebrate groups (Collembola, Coleoptera, Diptera, and Formicidae), we compared the number of that prey item in toads' stomachs from the two pH treatments. We tested whether the number of prey items was different between soil treatments using a nonparametric Kruskal-Wallis test on the average number per stomach for each pen. We conducted all analyses using R software environment version 3.6.1 (R Core Team, 2019).

RESULTS

Soil pH

Soil pH differed between lime-treated and untreated forest plots ($\chi^2 = 41.27$, P <0.001). Mean soil pH in lime-treated forest plots was 6.0 ± 0.1 s.e. compared to 4.4 ± 0.1 s.e. in untreated forest plots. These microhabitat pH values were consistent with measurements taken from each 800 m^2 plot in August 2017, when mean pH values in lime-treated and untreated forest plots were 5.9 ± 0.1 s.e. and 4.2 ± 0.1 s.e., respectively. There were no differences in pH among the enclosure treatments ($\chi^2 = 1.45$, $P = 0.48$).

Effects on American toad survival and growth

There was no effect of soil pH treatment ($\chi^2 = 0.37$, $P = 0.54$), forest ($\chi^2 = 1.23$, $P = 0.54$), or their interaction ($\chi^2 = 3.69$, $P = 0.16$) on toad survival to 90 days. We located an average of 38% of toads on day 90 (Figure 2A). Toad mass at the initiation of the study (i.e., mass at metamorphosis) did not differ between soil pH treatments ($F_{1,14} = 0.81$, $P = 0.38$) or among the three forests ($F_{2,14} = 1.45$, $P = 0.27$). Mean mass at metamorphosis was 0.29 g ± 0.01 s.e. for all enclosures. Toad growth (body mass measured after 25, 60, and 90 days) was not significantly affected by forest or soil pH treatment, although there was a tendency ($P = 0.06$) for toads in plots with elevated soil pH to be larger than toads in plots with acidic soil pH (Fig. 2B, Supplemental Table 1).

Effects on invertebrate abundances

Collembola were the most abundant invertebrates collected among all subplots, both in samples collected on day 0 and on day 91 (84% of all invertebrates collected, by number of individuals collected), followed by Diptera (5%), Acari (4%), Coleoptera (2%), and Araneae (2%). All other groups (Chilopoda, Diplopoda, Gastropoda,

Hemiptera, Hymenoptera, Isopoda, Lepidoptera, Oligochaeta, and Orthoptera) comprised <1% of total captures. There was a significant interaction between soil pH treatment and subplot type on Acari abundance on day 91 ($\chi^2 = 10.08$, $P = 0.01$). Mean Acari abundance in subplots with acidic soil pH and no constructed enclosure was 17.1 ± 5.8 s.e. compared to 5.4 ± 0.6 s.e. for all five other soil treatment and enclosure type combinations. There were no other significant interactions between or among soil pH treatment or subplot type on abundances of the most common invertebrate groups collected (Supplemental Table 2). There was also no main effect of soil pH treatment on the abundances of any of the invertebrate groups nor overall invertebrate diversity for all samples collected either on day 0 or on day 91 (Supplemental Table 2). NMDS and resulting ANOSIM further indicated no difference in invertebrate communities between the soil pH treatments on day 0 (NMDS, k = 2, stress = 0.036; ANOSIM R = 0.003, $P =$ 0.33) or on day 91 (NMDS, k = 2, stress = 0.054; ANOSIM R = -0.040, $P = 0.68$; Fig. 3).

There was a significant main effect of enclosure type on invertebrate abundances and overall taxonomic diversity in our linear mixed effect models. Post-hoc pairwise comparisons revealed a significant effect of enclosure presence ("no enclosure" treatment) on invertebrate abundances. We consistently collected higher abundances from most subplots without enclosures compared to those with enclosures (both with and without toads), both on day 0 and day 91 (Fig. 4). Post-hoc pairwise comparisons revealed no differences in initial invertebrate group abundances or overall diversity on day 0 between samples collected from enclosures with toads and those enclosures without toads. (Fig 4). After 91 days, only Araneae abundances differed between these two enclosure types; mean Araneae abundance was 3.8 ± 0.5 s.e. in enclosures without toads compared to 2.0 ± 0.4 s.e. from enclosures with toads. There were no differences between enclosures with and without toads for any other invertebrate group nor overall taxonomic diversity on day 91 (Fig. 4).

Effects of soil pH treatment and invertebrates on American toad diet

We evaluated the stomach contents of 27 toads surviving to day 90 (13 toads from forest plots with elevated soil pH and 14 toads from forest plots with acidified soils). The toad diets were similar, with a Pianka's index of niche overlap of 0.77. The four most important prey items in both groups of toads were Collembola, Formicidae, Diptera and Coleoptera (Table 1). There was no difference in number of prey in toad stomachs for any of these groups (Collembola ($\chi^2 = 0.34$, $P = 0.56$), Diptera ($\chi^2 = 0.09$, $P = 0.77$), Coleoptera ($\chi^2 = 3.15$, $P = 0.08$), or Formicidae ($\chi^2 = 3.26$, $P = 0.07$).

DISCUSSION

We used the differences observed between forest plots with elevated soil pH and those plots with acidic soil pH to assess how acidification has affected a common forest floor amphibian and its invertebrate prey. Although not statistically significant, there was a tendency throughout the study for toads in forest plots with elevated soil pH to be larger than toads growing on more acidic soils. After 90 days, toads from forest plots with experimentally elevated soil pH were 14% larger than toads developing on still-acidified soils. Growth immediately following metamorphosis is an important aspect of amphibian survival, both during the first season post-metamorphosis and for overwintering success (Boone, 2005; Distel and Boone, 2009). Thus, this difference in growth we observed could potentially translate to differential survival for first-year American toads in both the coming winter and the following growing season. Our invertebrate pitfall sampling results, however, provided no evidence that the differences we observed in American toad growth between forest plots with elevated soil pH and forest plots with acidified soils were mediated by a bottom-up effect of invertebrate prey availability. We found no negative effect of acidic soil pH on invertebrate abundances or overall diversity, nor any difference in invertebrate community structure between the two soil pH treatments. We also found no differences in the diet of American toads raised in these soil pH treatments based on examination of their gut contents following 90 days. Because prey was not limited for toads in enclosures with elevated soil pH compared to acidic soil pH, we do not find any support for the hypothesis that forest acidification indirectly affects growth of American toads due to a negative effect of acidified soils on prey availability.

American toad survival did not differ between soil pH treatments. The survival rates we observed are consistent with other studies that have held American toads at a starting density of four toads per 1 m^2 for approximately three months (Harper and Semlitsch, 2007). All but three of our enclosures had at least one toad survive to day 90. No toads survived to day 90 in one enclosure from a forest plot with elevated soil pH and in two enclosures from forest plots with acidified soils. Examination of the invertebrate samples from these enclosures did not provide any evidence that these locations were prey-limited and thus do not suggest that toad survival was affected by a lack of food. The two enclosure locations in forest plots with acidic soil pH where no toads survived had measured pH values below 4.0, and no toads were located in these enclosures on day 60 nor day 90. These were the only enclosures with a measured pH below 4.0 (the enclosure from a forest plot with elevated soil pH where no toads survived to day 90 had a pH of 5.8). These results suggest that toad survival may have been affected by this low pH, and this is supported by previous studies that have documented that American Toads are exclusively associated with soil pH values higher than 4.0 (Wyman, 1988). The distribution of other forest floor amphibians such as the Red-backed Salamander may also be limited by soil pH, especially in locations where pH is below 4.0 and where these acidified soils may physiologically affect the salamanders (Wyman and Hawksley-Lescault, 1987). However, recent studies have demonstrated that Red-backed Salamanders are locally adapted to acidified habitats and can tolerate pH values below 3.8 (Moore and Wyman, 2010; Bondi et al., 2016). Although our study did not explicitly test the physiological response of American toads to acidic soils, our results from these extremely acidic subplots and the lack of evidence for survival being limited due to prey availability support previous findings that soil pH below 4.0 may reduce amphibian survival via direct physiological stress.

We found no effects of acidified soils on invertebrate abundances at either the beginning or end of our study. Previous studies of invertebrate community responses to acidified soils suggest that these responses tend to be taxa-specific and not always consistent. Kuperman (1996) reported higher abundances of certain macroinvertebrates, including Diptera, Arachnids, and some Coleoptera, in oak-hickory forests of Illinois (average soil pH 6.5) compared to forests in Ohio and Indiana (average soil pH 4.5). However, abundances of other groups such as Lepidoptera and certain Coleoptera were higher in the more acidic forests of Ohio and Indiana. In these same forests, Collembola abundances were not affected by soil pH under normal climatic conditions, but a period of drought contributed to abundance reductions of up to 85% in the more acidified forests

(Kuperman et al., 2002). Chagnon et al. (2001) examined Collembola diversity in a sugar

maple forest of Quebec with acidified soil. The authors used experimental liming to

elevate soil pH in some locations, and total Collembola abundance was higher in those

plots with still-acidified soils compared to those with elevated soil pH. However, overall

Collembola diversity was lower in plots with still-acidified soils. pH values for untreated

locations in that study averaged 4.2 compared to 4.9 and 5.9 for elevated pH locations

treated with a low or high amount of lime, respectively. Those pH values are similar to

those measured in our study (4.3 and 5.9 in untreated, acidic soils and elevated pH soils,

respectively), yet we documented no effect of acidified soils on Collembola abundances.

Fisk et al. (2006) documented no effect of acidified soils on Collembola abundances or

overall diversity compared to those soils with experimentally elevated soil pH. In the

same study, Acari abundances and diversity were higher in forests locations with

acidified soils compared those with elevated pH, although the changes to forest floor pH

via experimental Hi-Ca additions and overall soil pH values studied by Fisk et al. (2006)

were generally smaller (approximately 3.6 to 4.3 in untreated and treated forest sites,

respectively, after three years) than those we studied. Higher Acari abundances and

overall diversity in acidified soils have also been documented in in several other studies

(Hågvar and Amundson, 1981; Huhta et al., 1986). Although we did not specifically

examine diversity responses (e.g., genus- or species-level responses) within the

invertebrate groups we collected and thus cannot determine if these were affected by

acidified soils, we found no evidence that acidified soils negatively affect the abundances

of any of the invertebrate groups we collected. However, we acknowledge that our use of pitfall traps exclusively to sample invertebrates, instead of in combination with other methods such as Berlese extraction funnels, may have limited our ability to measure abundances of certain invertebrates. This may be true for certain groups, such as Acari, that are likely to be more active within the interstitial spaces of the forest floor instead of on the soil surface.

Collembola were by far the most abundant invertebrate group collected from our pitfall traps at the start of the study and after 91 days. However, Collembola made up only 16% of diets recovered from American toads after 90 days. Collembola are a common prey item for *Anaxyrus* toads immediately following metamorphosis, but toad diet changes throughout ontogeny due to gape-limitations, especially during the first season following metamorphosis as preferences tend to shift from Collembola and Acari to larger Coleoptera and Hymenoptera as toads grow (Flowers and Graves, 1995). This shift in diet may account for the distribution of diet items we recovered from toads collected after 90 days. Toads in our study demonstrated a generalist diet based on the invertebrate diversity we collected from the stomachs of our 27 surviving toads on day 90. Bondi et al. (2019) suggest that a generalist feeding strategy may explain why Red-backed Salamanders are able to tolerate soils with pH values below 3.8. Similar to other diet generalists such as the Red-backed Salamander, we suggest that a generalist feeding strategy in American toads may explain, in part, why we documented no significant difference in toad survival or growth between the two soil pH treatments in our study.

Amphibians serve an important function in forest ecosystems through their potential roles in invertebrate regulation, nutrient cycling, and soil dynamics (Davic and Welsh Jr., 2004; Hocking and Babbitt, 2014a). They are often abundant and comprise a large proportion of the total biomass present; in some locations, salamander densities can reach as high as 12,900 inidividuals·ha^{-1} (Burton and Likens, 1975; Semlitsch et al., 2014). As a key component in forest floor trophic dynamics, some experimental evidence suggests that amphibians may play a large role in suppressing abundances of common invertebrate groups such as Coleoptera, Diptera, and Araneae (Wyman, 1998). However, other studies have documented no effects of forest floor amphibians on invertebrate communities or terrestrial ecosystem functioning (Homyack et al., 2010; Hocking and Babbitt, 2014b). Therefore, the importance of amphibians in regulating forest floor communities is still an open area of investigation. Comparing invertebrates in enclosures with toads to enclosures without toads, we only found a difference in the abundances of Araneae collected on day 91. We collected more Araneae from enclosures without toads compared to enclosures with toads. However, only two of the 27 toads that survived to day 90 had Araneae in their stomachs (one per toad). This suggests that Araneae do not make up a substantial portion of the toad diet at this stage of ontogeny, which is comparable to other diet studies of the genus *Anaxyrus* (Flowers and Graves, 1995). We found no other effect of toad presence on the other invertebrate groups. Thus, amphibian control of invertebrate populations may not always constitute a clear top-down mechanism in forest floor trophic dynamics. Instead, we documented a positive correlation between toad mass and Collembola abundances we collected on day 91 regardless of soil pH. Invertebrate abundances, especially Collembola, may therefore influence forest floor amphibians such as American toads in a bottom-up manner in this system, instead of amphibians exerting a top-down effect on the forest floor invertebrate community. Trophic systems are typically affected by a combination of top-down and bottom-up effects (Ponsard et al., 2000), so the effect of forest floor amphibians on ecosystem function is likely context dependent.

We documented a strong effect of constructed enclosures on invertebrate abundances and overall invertebrate diversity. There were consistently higher abundances of all common invertebrate groupings in pitfall trap samples from non-enclosure subplots compared to subplots with constructed enclosures. In some groups, such as Araneae and Coleoptera, there were nearly nine times as many individuals collected at subplots without enclosures compared to subplots with enclosures. Overall diversity was also consistently greater both on day 0 and day 91 in samples collected from non-enclosure subplots. Enclosures were constructed between 26 and 43 days before invertebrates were sampled on day 0, suggesting that even after as few as four weeks these enclosures can significantly affect the forest floor invertebrate community. After an additional 90 days we saw continued enclosure effects on the invertebrate community. We propose that the enclosures likely modified the microhabitat characteristics of the forest floor via changes in temperature and humidity, and this may negatively influence the forest floor invertebrate community. The structures also may have inhibited the invertebrate groups we examined from moving within and among of the 1 m^2 subplots, thereby both reducing our capture rates within enclosures and reducing the ability of invertebrates to recolonize those areas of the forest floor where enclosures were present. These effects are especially important to consider because in situ terrestrial enclosures such as those we constructed are often used in conservation studies examining the responses of amphibians to management practices (e.g., Earl and Semlitsch, 2015). When terrestrial enclosures strongly influence the invertebrate communities as we have documented, this becomes an important study design consideration for research conducted by forest ecologists and managers. Terrestrial enclosures are also used in a variety of other amphibian ecological studies, including those focused on density dependence, habitat selection, and responses to contaminants (Harper et al., 2010). Therefore, consideration of the effect of these enclosures on invertebrate communities should be a consideration for any future study utilizing these structures.

Hardwood forests in temperate regions of the world can be especially sensitive to acidification, and understanding the ecological implications of acidification on specific taxa as well as trophic interactions are often important goals for ecologists, conservation biologists, and land managers. We found that acidic forest soils may have the potential to negatively affect American toads following metamorphosis via reduction in growth. However, this was not caused by an indirect effect of acidification on prey availability. Instead, the potential for negative effects of acidified soils on American toad growth may be due to direct physiological effects, as have been documented for other forest floor amphibians. Forests with experimentally elevated soil pH can be used to mimic those conditions prior to historic acidification. We encourage future in situ research in acidified forest systems to employ methods to experimentally manipulate soil pH in order to further examine the potential for both direct and indirect effects of acidification on American toads and other forest floor amphibians. Continued research should also focus on the potential for intraspecific tolerance to acidity in not only diet generalists such as the American toad and Red-backed Salamander but more specialized species as well.

TABLES

Table 1. Importance (I_x), numerical proportion of prey items (n_x/N), and relative frequency (f_x/F) of prey items in toad stomach from pens with elevated soil pH and pens with acidified soil.

Prey type	Elevated soil pH			Acidified soil		
	I_x	n_x/N	f_x/F	I_x	n_x/N	f_x/F
Collembola	0.21	0.24	0.18	0.14	0.15	0.12
Coleoptera	0.11	0.008	0.14	0.21	0.23	0.20
Diptera	0.20	0.25	0.16	0.15	0.14	0.16
Formicidae	0.25	0.27	0.23	0.11	0.12	0.10
Acari	0.07	0.04	0.09	0.11	0.10	0.12
Hymenoptera (non-Formicidae)	0.07	0.06	0.07	0.10	0.08	0.12
Diplopoda	0.02	0.01	0.02	0.04	0.03	0.06
Orthoptera	0.00	0.00	0.00	0.05	0.05	0.06
Hemiptera	0.03	0.02	0.05	0.06	0.08	0.04
Isopoda	0.02	0.01	0.02	0.00	0.00	0.00
Arachnid	0.02	0.01	0.02	0.01	0.01	0.02
Chilopoda	0.00	0.00	0.00	0.01	0.01	0.02

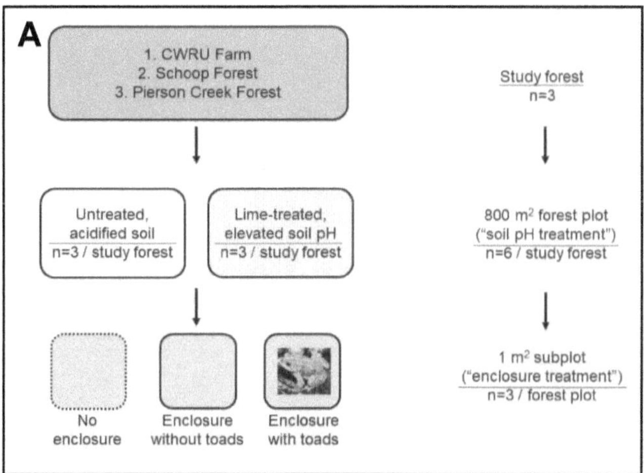

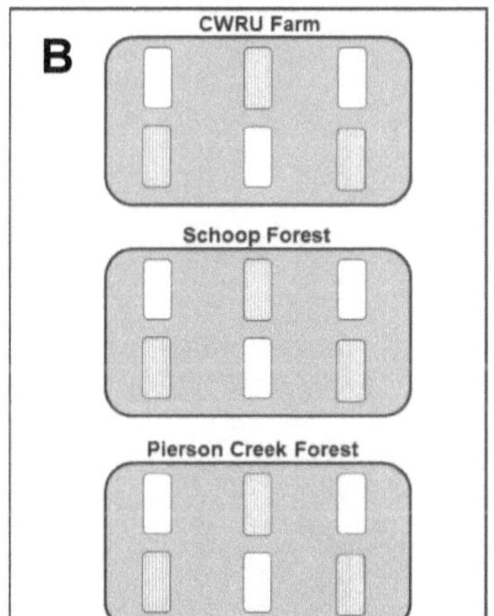

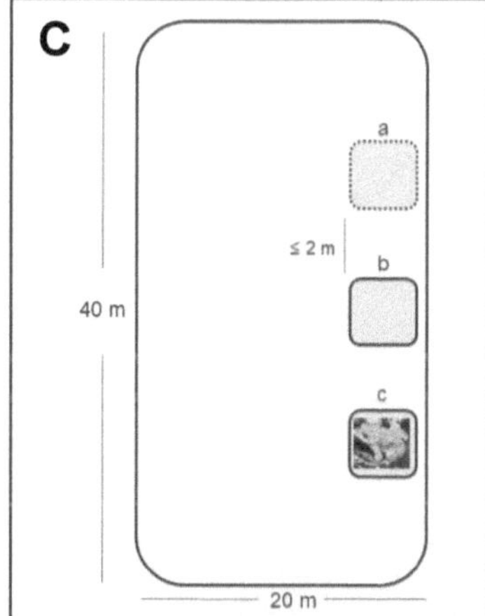

Figure 1. Schematic demonstrating the experimental design, location of forest plots within the three study forests, and location of 1 m² subplots within a forest plot. A). General layout of experiment among our three scales of interest: study forest, forest plot,

and subplot B). Typical arrangement of six forest plots within each of the three study forests. Within each forest, white rectangles represent the three forest plots with untreated, acidified soils while lined rectangles represent the three lime-treated, elevated soil pH forest plots. C). An enlarged example from (B) of one possible random arrangement of the three $1m^2$ subplots (light gray boxes; not drawn to scale) located along the perimeter of a forest plot with untreated, acidified soil. Each $1m^2$ subplot in each forest plot was assigned to one of three enclosure treatments: a) no enclosure, b) enclosure without American toads, and c) enclosure with four American toads.

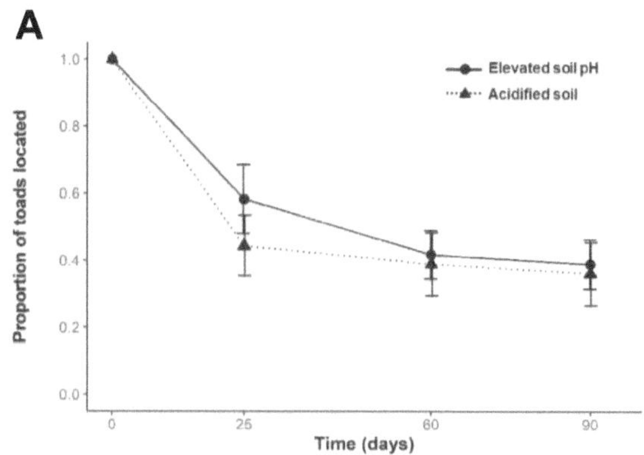

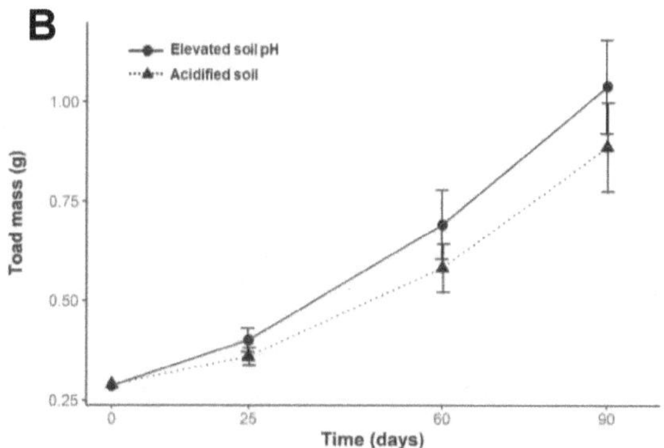

Figure 2. American toad survival (proportion of initial toads, n=4 per enclosure, located) and body mass over 90 days within enclosures located in forest plots with elevated soil pH (circles and solid line) and forest plots with acidified soils (triangles and dashed line). A). Proportion of American toads located in each enclosure within each soil pH treatment after 25, 60, and 90 days. Data are mean per soil pH treatment at each census period ± 1 s.e. B). American toad mass in each enclosure within each soil pH treatment at the start of the study and after 25, 60, and 90 days. Mean toad mass was calculated per enclosure at each census period. Data are mean per soil pH treatment at each census period ± 1 s.e.

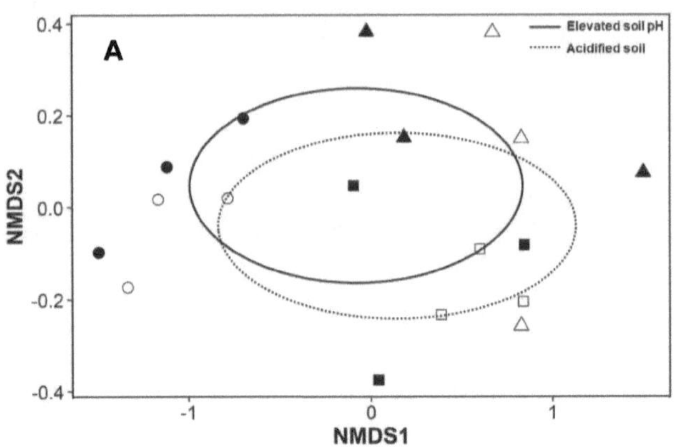

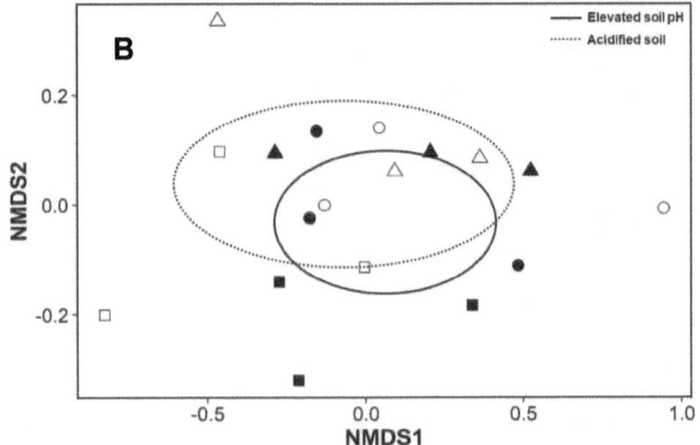

Figure 3. Non-metric multidimensional scaling (NMDS) ordination of invertebrate community composition on day 0 (A) and day 91 (B). Soil pH treatment group scores and standard errors are represented by symbol type and line type (sites with elevated soil pH – solid shapes, solid line; sites with acidified soils – open shapes, dashed line). Forests are represented by symbol shape (CWRU Farm – circle; Schoop Forest – triangle; Pierson Creek Forest – square).

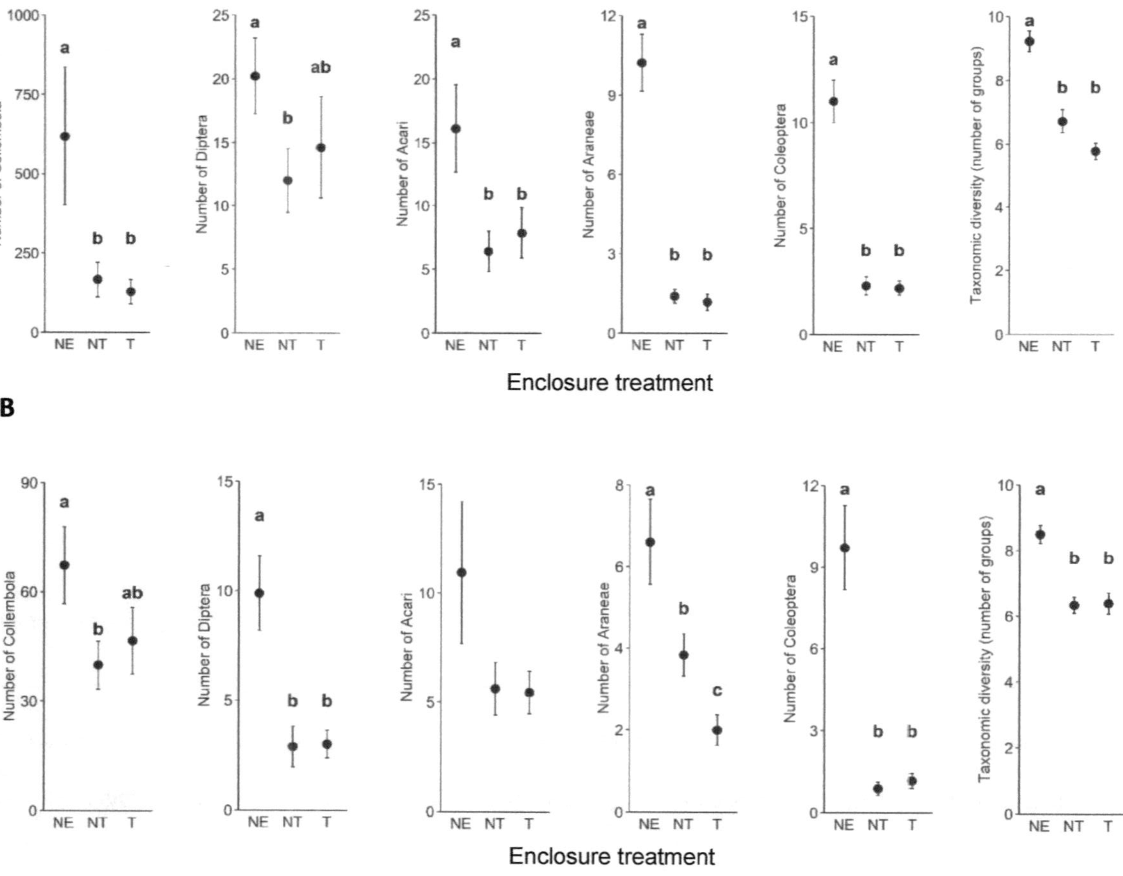
A
Number of Collembola
Number of Diptera
Number of Acari
Number of Araneae
Number of Coleoptera
Taxonomic diversity (number of groups)
Enclosure treatment
NE NT T
B
Number of Collembola
Number of Diptera
Number of Acari
Number of Araneae
Number of Coleoptera
Taxonomic diversity (number of groups)
Enclosure treatment
NE NT T

Figure 4. Invertebrate abundances collected on day 0 (A) and day 91 (B) from subplots with three different enclosure treatments (NE = No Enclosures: no enclosures present; NT = No Toads: enclosures present and no toads; T = Toads: enclosures present with toads). Data are mean ± 1 s.e. For each invertebrate grouping where the overall model was significant, enclosure types with the same letter were not different at $P < 0.05$ from post-hoc tests.

SUPPLEMENTAL TABLES

Supplemental Table 1. Summary of repeated measures ANOVA for effects of forest stand and soil pH treatment on American Toad body mass over the 90-day study period.

	Source of variation	Df	F	*P*
Between	Forest stand	2	2.36	0.10
	Soil pH treatment	1	3.90	0.06
Within subjects	Time	1	24.37	<0.001
	Forest stand x Soil pH treatment	2	0.12	0.89
	Forest stand x Time	2	1.38	0.26
	Soil pH treatment x Time	1	1.88	0.18
	Forest stand x Soil pH treatment x Time	2	0.13	0.88
	Error	50		

Supplemental Table 2. Linear mixed-effects model results for soil pH treatment, enclosure type, and their interactions on invertebrate abundances and overall invertebrate diversity collected on day 0 and day 91.

Invertebrate group	Day		χ^2	P
Collembola	0	Soil pH treatment	0.537	0.464
		Enclosure type	33.054	**<0.001**
		Soil pH x enclosure	0.675	0.713
	91	Soil pH treatment	0.465	0.496
		Enclosure type	6.714	**0.035**
		Soil pH x enclosure	0.606	0.739
Diptera	0	Soil pH treatment	0.992	0.319
		Enclosure type	17.102	**<0.001**
		Soil pH x enclosure	0.473	0.789
	91	Soil pH treatment	0.001	0.978
		Enclosure type	46.594	**<0.001**
		Soil pH x enclosure	0.951	0.621
Acari	0	Soil pH treatment	0.223	0.637
		Enclosure type	18.280	**<0.001**
		Soil pH x enclosure	0.708	0.702
	91	Soil pH treatment	0.732	0.392
		Enclosure type	2.068	0.356
		Soil pH x enclosure	10.079	**0.006**
Araneae	0	Soil pH treatment	0.396	0.529
		Enclosure type	158.341	**<0.001**
		Soil pH x enclosure	9.962	**0.007**
	91	Soil pH treatment	0.324	0.569
		Enclosure type	46.541	**<0.001**
		Soil pH x enclosure	5.786	0.055
Coleoptera	0	Soil pH treatment	0.018	0.892
		Enclosure type	102.318	**<0.001**
		Soil pH x enclosure	0.523	0.770
	91	Soil pH treatment	0.832	0.362
		Enclosure type	58.254	**<0.001**
		Soil pH x enclosure	0.094	0.954
Overall Diversity	0	Soil pH treatment	0.144	0.704
		Enclosure type	59.079	**<0.001**
		Soil pH x enclosure	1.260	0.533
	91	Soil pH treatment	1.288	0.257
		Enclosure type	32.824	**<0.001**
		Soil pH x enclosure	1.828	0.401

Conclusion

Advancements of my research

I investigated how biotic and abiotic characteristics affect habitat selection, and the effects of using those habitats, on two common North American amphibians. My goal was to better understand how and why amphibians use the habitats they do and what the consequences are of using those habitats. I was especially interested in examining these consequences among different amphibian life stages. I found that biotic and abiotic characteristics of habitats can affect amphibian development and male and female use of breeding habitats in complex ways. These effects have implications for individual fitness, population dynamics, and community assembly. This can, in turn, be important for both conservation and management decisions when habitats are created, modified, or restored, thus altering the biotic and abiotic characteristics as well as spatial locations of available habitats on the landscape.

Habitat selection and breeding site selection are important for an individual's fitness and that of their offspring (Morris 2003, Schmidt et al. 2010). Costs and constraints must be accounted for by an animal when determining if it should use a certain habitat over another (Acker et al. 2017). My use of a breeding site choice experiment in the field demonstrated that female eastern gray treefrogs select breeding habitats without heterospecific larvae present to avoid future competition for their developing offspring. This study fills an important gap in the current literature examining how nonpredatory heterospecific competitors affect the female preference-offspring performance theory in habitat selection. My research demonstrates that female breeding site selection is an adaptive behavior for offspring performance in the presence of nonpredatory heterospecific competitors, an area of research that is lacking in habitat selection studies. Relatively few studies that have tested if anurans prefer to breed in habitats based on the presence of heterospecific anurans, and these past studies have produced mixed results including attraction to, avoidance of, and no preference between habitats with and without heterospecifics (Resetarits and Wilbur 1989, Indermaur et al. 2010, Dananay and Benard 2018). Tests of whether anuran preference for habitats is correlated with offspring performance are less common, and these studies have also largely focused on predator and conspecific effects on resulting offspring performance (Rieger et al 2004, Pintar and Resetarits 2017, Buxton and Sperry 2017). Very few studies have experimentally investigated the female preference-offspring performance as I did in the presence of heterospecific competitors. Continued research in this area will help fill this gap.

Males and females may use similar or different cues to assess habitat quality when selecting a breeding site (Eterovick and Ferreira 2008). Whether or not male and female responses to habitat quality are similar and coordinated with one another or different and in conflict with one another is important for understanding why individuals use the habitats they do during the breeding season. Although much is known about how male signaling affects female mate choice and how a male's signal indicates his quality as a potential mate, the use of habitats of varying quality by males in a given population producing signals of different quality is an understudied area of habitat selection research. My study assessing which breeding habitat types male treefrogs use among those with and without larval competitors present expanded upon previous research (e.g., Resetarits and Wilbur 1991) by also characterizing the quality of these males as parents based upon their advertisement calls. These types of studies combining both male signaling traits and use of habitats of varying quality are rare. By documenting that, at least in the absence of other competing males, males calling from certain habitats in our study are not inferior to those at other habitats in their signaling, we would not expect females breeding at their preferred habitats (those without larval competitors) to be mating with males of any certain quality. Therefore, female preference for habitats that benefit offspring performance may be more important than the quality of the male calling at available habitats. My follow-up study in Chapter 3 comparing fringe habitat males to core habitat male indicated that males using these fringe habitats are of similar size and quality (as measured by call characters) to those males using the core habitat. Because males of similar size and quality use habitats of both varying quality (presence or absence of larval competitors) and location (fringe or core) similarly, factors besides body size and individual male quality are influencing their use of these habitats. Instead, use of these fringe habitats by males in both Chapter 2 and Chapter 3 provide evidence of signaling plasticity that may allow a male to remain active for more nights throughout the breeding season. This is important because the number of nights spent advertising is a better predictor of male mating success than advertisement call characters or body condition in many species of frogs, including *H. versicolor* (Gerhardt et al. 1987; Sullivan and Henshaw 1992, Bertram et al. 1996).

Acidification and resulting changes to soil pH is an abiotic condition that can have broad effects on many habitats. While the effects of acidification on certain aspects of forest ecosystems have been well studied, less is known about the influence of soil acidification on the forest floor food web that includes amphibians and invertebrates. I found only a marginal negative effect of acidified habitats on American toad growth, but neither toad survival nor diet were affected. I also found no effect of pH on the invertebrate community or forest floor trophic dynamics. However, the presence of temporary enclosures we constructed in our study significantly reduced invertebrate abundances and overall diversity. When terrestrial enclosures strongly influence the invertebrate communities as we have documented, this becomes an important study design consideration for research and applied practices conducted by ecologists and managers.

Future directions

Although there is a myriad of biotic and abiotic conditions that affect amphibians and other organisms, understanding the role of these characteristics in habitat selection and use in heterogeneous landscapes can play a critical role for individual fitness, population dynamics, and community structure. These biotic and abiotic characteristics can affect habitat selection and use in complex ways, resulting in differences within individuals of a given species, between males and females, and across life stages. These effects can have important implications for other taxa beyond amphibians, and increasing our knowledge of these types of relationships in ecological applications can provide insight into best practices for land managers and conservationists.

Female eastern gray treefrog supported the "mother knows best" hypothesis in my study, demonstrating that they select breeding habitats without two heterospecific anuran larvae present to avoid future competition for their developing offspring. Very few studies have experimentally investigated the female preference-offspring performance as I did, and this is important not only in the context of other potential competitor species but other biotic and abiotic conditions as well. Offspring performance does not always match female site preference in other organisms, as females may select habitats which increase their own longevity while reducing their offspring survival (i.e., the "optimal bad motherhood" principle; Garcia-Robledo and Horvitz 2012). Additional possibilities of maladaptive breeding behavior for offspring can arise due to ecological traps resulting from anthropogenic disturbances and created habitats (Schlaepfer et al. 2002).

As I have demonstrated, male use of breeding habitat quality cues may not always completely align with those of females. Furthermore, males calling from habitats of perceived different quality to females for egg deposition did not demonstrate any difference in the advertisement calls they produced at these habitats of varying quality. This may have been due to the lack of competing males in the immediate vicinity of the focal males. The social environment surrounding an advertising male affects his call characters. Future research should focus on moving males from these isolated social environments to a competitive interaction involving one or more additional calling males. The call characters in a more competitive social environment could then be compared to these baseline call characters of isolated males at these habitats with heterospecifics present or absent. This would allow a better assessment of if males using habitats of different quality produce calls of different quality once in a competitive social context. My results, however, do confirm that these fringe males will adjust their advertisement call based on the social environment, as demonstrated when I moved fringe males to the core breeding habitat and assessed the changes to their calls. However, I cannot say with certainty if males using the fringe habitat differed in body size or body condition compared to the core habitat population because my sample sizes of males from both the fringe and core habitats were low. Future work will continue in this study system to increase the number of fringe males that can be tested as well as the number of males collected at the core habitat for size measurements. A more robust sample size of frog body mass and body condition within both resident types will allow for a better comparison of males using these two habitat types. This will help in our understanding of if there is a phenotypic difference between males using fringe habitats for breeding compared to males utilizing the core breeding pond.

Although American toads were only marginally affected by acidified soils, this may be due to local adaptation among the populations we studied, as has been demonstrated for other amphibians on extremely acidified soils (Bondi et al. 2016, Bondi et al. 2019). Our source populations of American toads came from ponds in historically acidified habitats, which may have contributed to our results. Additionally, there is the possibility for carry-over effects between life stages in amphibians due to their complex life histories. Additional research should examine how toad larvae raised in either acidified or more neutral aquatic habitats then survive and develop following metamorphosis in both acidified and more neutral forests.